WA WOM

THE COMING AGE OF SOLAR ENERGY

Books by D. S. Halacy, Jr.

THE COMING AGE OF SOLAR ENERGY, REVISED EDITION

THE GEOMETRY OF HUNGER

MAN AND MEMORY

COMPUTERS: THE MACHINES WE THINK WITH

THE WEATHER CHANGERS

CYBORG: EVOLUTION OF THE SUPERMAN

THE COMING AGE OF

HARPER & ROW, PUBLISHERS

SOLAR ENERGY

Revised Edition

D. S. HALACY, JR.

New York, Evanston, San Francisco, London

 For information address Harper & Row, Publishers, Inc., 10 East 53rd Street, New York, N.Y. 10022. Published simultaneously in Canada by Fitzhenry & Whiteside Limited, Toronto.

STANDARD BOOK NUMBER: 06–011714–1

LIBRARY OF CONGRESS CATALOG CARD NUMBER: 72–79670

CONTENTS

1.	Smoke and Ashes: Heritage of Power	1
2.	A Look at the Sun	19
3.	The Solar Pioneers	34
4.	We Discover the Sun—Again	52
5.	Solar Power for the Space Age	75
6.	Putting the Sun to Work	97
7.	The White Magic of Photochemistry	118
8.	Electric Power from the Sun	134
9.	The Orbiting Solar Power Plant	149
10.	Sea Thermal Energy	167
11.	A Billion Kilowatts of Sunshine!	189
12.	Sun in Our Future	207
	Further Reading	221
	Index	225

"Power corrupts" was written of man's control over other men but it also applies to his control of energy resources. The more power an industrial society disposes of, the more it wants. The more power we use, the more we shape our cities and mold our economic and social institutions to be dependent on the application of power and the consumption of energy. We could not now make any major move toward a lower per capita energy consumption without severe economic dislocation, and certainly the struggle of people in less developed nations toward somewhat similar energy-consumption levels cannot be thwarted without prolonging mass human suffering. Yet there is going to have to be some leveling off in the energy demands of industrial societies. Countries such as the U.S. have already come up against constraints dictated by the availability of resources and by damage to the environment.

Earl Cook, *Scientific American* (September 1971)

1 SMOKE AND ASHES: HERITAGE OF POWER

Among the many problems facing us today are two crises that are more closely linked than is generally realized. On the one hand—and very obviously—we are beginning to choke on the wastes of our machine age. On the other—and not as evident to most of us, but nonetheless true—we are within sight of the end of fuel supplies for those machines which are the causes of environmental pollution. Shaping up in our future is the legendary meeting of irresistible force and immovable object. We continue to demand energy and more energy for the technological life we lead; in so doing we foul the air, earth, and water while we reach ever nearer the bottom of the fuel bin.

Most of mankind is power mad, in the sense that energy consumption is the ultimate status symbol, the gauge of the affluent

FIG. 1. Pollution: the inevitable result of burning fuel to produce power.

society. Energy consumption produces the gross national product (GNP), and the conspicuousness of that consumption is evident in the scarred landscape, the blighted atmosphere, and the fouled waterways. At times it seems that only the exhaustion of fuel will halt our desecration of the environment.

In a kind of desperate joy we have turned to nuclear energy as a follow-up for fossil fuels soon to be exhausted. Yet nuclear power poses enough of an environmental threat that strong pressures are slowing its exploitation. The last hope of the avid consumer of energy is held to be the development of fusion power, a peaceful harnessing of the power of its H-bomb. After more than two decades of sophisticated research involving the expenditure of billions of dollars, however, success is still around some distant corner. And in the opinion of some experts that corner will never be turned.

As it happens, there is an H-bomb we might use to solve both these pressing problems, an H-bomb that is presently bombarding the Earth with the equivalent of four thousand horsepower per acre of land surface. At 93 million miles it is far enough away not to present the hazards of an Earth-based nuclear plant. The sun, on which our lives have always depended, could do far more for us. But, like fish unaware of the water they swim in, few of us grasp the energy potential at the center of our solar system.

Hundreds of years ago men demonstrated that sunshine was good for more than growing crops and tanning our skin. A long line of engines, heaters, distillation plants, and electric-power converters have been constructed, and a handful are in token operation. Yet for some reason we have hesitated to seize this opportunity which is so evident each time a new day dawns.

Had we been brought up in a technology that made use of solar energy, we might well reject with horror any suggestion that we switch to internal-combustion engines, or nuclear reactions that

FIG. 2. Desecration caused by strip mining.

produce harmful radiations as by-products. In the reverse situation that prevails we perversely resist the substitution of solar energy for fossil fuel consumption. There are hints, however, that at last we may be more willing to accept the sun's offering.

Up to now the expenditure for solar energy research and development has been practically nil, and most of the work in this direction has been done by independent researchers. For years these solar prophets have pleaded, cajoled, and threatened. They have also blueprinted solutions to our problems. One exciting proposal suggests an orbiting solar collector that would beam energy to Earth as microwaves. Another envisions a huge solar-power complex in the Southwest desert country, producing sufficient electricity for all of the United States. A third promising idea is the "sea thermal energy" plant, in effect a steam engine tapping the nearly limitless heat energy of the oceans. These are but a few of the multitude of potential applications for solar energy. And at last, in the face of serious environmental problems, they are being taken more seriously. The time seems ever more propitious for the coming of age of the oldest source of energy there is.

Quest for Power

When our country was formed, a couple of hundred years ago, most of the fuel burned was wood. This accounted for 75 percent of the need. For the rest man depended on wind and water and the muscles of animals and men. Examine the old mills at Sturbridge, Massachusetts, and elsewhere, and recall that the mill towns in New England were built on rivers where waterpower could turn wheels to operate machinery. But these days are gone forever. It has been pointed out that all the windmills in the country produce about as much power as we use to operate electric fans. The total waterpower of the country takes care of only a small fraction of our needs. Instead of 4 million people in America, as there were in 1776, there are now about 210 million. And each of them needs—or has been convinced that he needs—

the equivalent of many horsepower in various kinds of machines, from automobiles to jet planes to air-conditioners.

Once begun, man's march from a primitive energy consumption economy to the modern sophistication we enjoy or deplore today came about quickly. By 1830 the increase in population and the increase in demand for power had boomed the consumption of coal, although as late as 1870 we still used three times as much wood as coal. However, steam locomotives (they opened the West, remember?), cotton gins, factories, mechanized reapers for farms, Stanley Steamers, and other machines were convincing mankind that they represented progress.

By the turn of the century coal had just about displaced wood as a fuel. Only the poor heated their homes with wood (today, ironically, only the rich can afford to!). The iron and steel industry consumed great quantities of coke; lamps burned "coal oil" or

FIG. 3. Gulf Coast oil tank blazes after damage caused by hurricane.

kerosene. And the petrochemical industry, including chemicals, dyes, and so on, also required vast stores of coal.

Natural gas was known for a long time before piping was begun to remote markets in about the 1870s. By 1930 it was being transported for thousands of miles, and after World War II it was a major contributor to our energy economy.

Petroleum, although known centuries ago and used to a slight extent, became a major source of energy only within the last century. In 1859 America's first oil well was brought in, and by the turn of the century the internal-combustion engine and the automobile began to demand huge amounts of gasoline and other petroleum products. Today half of all the petroleum used is for our transportation needs.

Electricity had been known for centuries, mostly as a curiosity. Battery-powered electric automobiles antedated the gasoline-powered version, but lost favor against the greater acceleration and economy of the gas buggies. Yet electricity was such a handy form of power that its use was inevitable. Edison's electric light spurred progress, and from 1883 on—and, especially after 1890, when the hydroelectric plant was developed—electric power plants have garnered an increasing share of the energy market. This clean source of power is unfortunately not available in very large quantities.

In his presidential address to the British Association in 1911, Sir William Ramsay expressed concern over increasing consumption of energy, pointing out that, ultimately, energy was limited. Ramsay mentioned the fact, too, that every British family then had an average of twenty energy slaves—he called them "helots"—working for them. Today each of us in the United States draws on the services of the equivalent of about seventy mechanical slaves to enjoy the good life. Raising the living standards of the poor would increase the figure to more than one hundred slaves, each consuming many kilowatt-hours of energy.

The first engines man invented, the waterwheels that came into being about a century before Christ, produced about one-half horsepower. By the seventeenth century waterwheels were still the most powerful engines in the world, and the famous Versailles "waterworks" at Marly-la-Machine, a showpiece for a long period of time, produced about seventy horsepower.

Today an engine of this power drives a "compact" car and is a very modest power source. Some automobiles have several hundred horsepower, aircraft engines deliver thousands of horsepower (although they are generally rated in pounds of thrust), and a rocket engine for spacecraft may produce more than 20 million horsepower! Electric power plants generate millions of horsepower.

As yet we have not looked at the whole picture of our dwindling stockpiles; these materials are important for other reasons than

FIG. 4. A family this size uses this much petroleum in a year.

their use merely as fuels. As an example, various processes in industry make use of carbon and the chemicals we are prone to burn up to make power. Thus, even if we have hundreds of years of fuel potential in coal and oil, we might do well to save it, or some part of it, for other uses than heat or power.

There are other shortages, too. Fresh water supplies are scarce in many parts of the world, and surely we must have drinking water whether our automobiles run or not. Heavy use of fuel to convert salty or otherwise undrinkable water to fresh will make further demands on the supply. Solar energy has proved itself in distillation plants, though as yet the costs of capital investment are not competitive with "cheap"-power distillation plants. Yet a scientist speaking at a symposium on the distillation of water once saw fit to mention that there was little reason to seek out ways of converting water using unconventional sources of energy: we seem to have enough fuel for *another century!* More than a decade later this shortsighted view still seems to prevail.

We can live a bit longer without food than without water, but we need food, too, more than we do fuel. Although the United States is now plagued with surpluses, most of the world suffers from too little food. If it strains us to feed 3.5 billion, what can we offer the 7 billion who may be hungry by the year 2000? Again we seem to be headed toward a Malthusian impasse. Nature did not leave us any stockpile of food; we must live on the income she provides. Some authorities see a situation eventually so desperate that man will be forced to till every available foot of soil, even farming in summer some areas he must leave in winter because of cold. Yet even with exhaustive use of the land, Earth can support just so many people, a density perhaps only twice that of present-day Japan.

Faced with an emptying fuel bin, what can man do about it? He could, of course, cut down on his consumption of energy. Life might be livable without inefficient two-ton automobiles, supersonic transportation, power lawn mowers, and electric toothbrushes. Some people do live with a minimum of mechanical contrivances. Surely, if it is impossible to produce more power, man must do without it. But all indications are that he will bend every effort, not only to maintain the present luxurious use of power, but even to increase it. This increase will be on two levels, also, when the underdeveloped peoples finally begin to catch up with their affluent brothers. One prediction foresees the use of fifty times as much energy a century from today.

No Turning Back

In 1970 the United States consumed 69×10^{15} Btus (British thermal units) of energy of which 95.9 percent was fossil fuel, 3.8 percent hydroelectric power, and 0.3 percent nuclear energy. Per capita consumption was 340 million Btus, equivalent to about 13 tons of coal or 2,700 gallons of gasoline. Of oil alone we are presently using a total of about 15 million barrels a *day*, a quantity which may be better appreciated by visualizing the string of 62,000 tank cars needed to carry that much fuel. In a decade the demand will probably double, running our consumption up to about 30 million barrels a day and 124,000 tank cars—which is fine if you are selling oil or tank cars. Even with our own wells running full blast, we still find it necessary to import about 25 percent of the oil we use, which leaves us dependent, not only on foreign producers, but on global political, economical, and military situations as well.

The United States presently consumes about 35 percent of the world's power. By the year 2000 this portion is expected to decline to only 25 percent—not because we will be curbing our demands for power, but because the rest of the developing world will be demanding relatively more. The rest of the world uses, on the average, only about one-fifth as much energy per capita as we do. By 2000 this may have jumped to about one-third.

Much as the fact disturbs conservationists, there is nevertheless a direct correlation between rate of energy consumption and a nation's material standard of living. It is easy to argue against this by saying that such a correlation presupposes that high energy consumption constitutes a high standard of living and that this is not necessarily true. However, the fact remains that countries such as India, far down the scale in per capita consumption of energy, are just as far down the scale in development. It is difficult to argue that these underdeveloped peoples have as high a standard of living as people in Sweden, Canada, or the United States.

How Much Fuel in the Stockpile?

As man used ever-increasing amounts of energy, engineers came up with a handy measure called the "Q." The Q is equal to 1 billion billion British thermal units (you might check the Btu rating of your furnace, which will probably be in the range of 100,000 Btus). One Q represents a lot of energy, and is equivalent to burning about 38 billion short tons of bituminous coal (the smoky kind).

One authority estimated in 1953 that world reserves of fossil fuels totaled about 27 Q, and that nuclear reserves might amount to about 575 Q. Here is a total of something over 600 Q, with the world presently using about 0.1 Q per year. This would seem to leave us six thousand years of power in the bank. The picture is not quite that straightforward, however. From the time of Christ to the year 1850 the world used up an estimated 9 Q of energy. In the next century we used up 4 Q, almost half that much. But predictions for energy consumption from 1950 to 2050 range as high as 487 Q! This would leave only a little more than 100 Q in the world fuel bank.

More recently, M. King Hubbert of Shell Oil has estimated even more optimistically that the initial supply of fossil fuels in the earth amounted to about 243 Q. Nuclear fuels may have amounted to "a few orders of magnitude" more than this. Three orders of magnitude would be a factor of 1,000; there would then be about 243,000 Q of nuclear fuel available to us, and our worries would seem to be over.

Unfortunately, even the most optimistic estimate of additional fuel reserves is not justification for continued rapid consumption. As the miners of so valuable a metal as gold have learned, there is a point at which it becomes uneconomical to produce a product. The President's Materials Commission in 1952 expressed this feeling about fuel reserves:

> The reserves of fuels with which to meet energy demands are relative. It is not a question of emptying the bin. We shall never do that. It is a question of how far down it is worth reaching in terms of economic cost. Such a question has no exact answer.

Nuclear Power: Fading Promise and Growing Threat

It has been learned the hard way that the rosy future painted for nuclear power plants was something of a mirage. Recently the AEC has been predicting about 170,000 megawatts of nuclear electric power in service by 1980. Now it seems that only about 100,000 megawatts will be available by that time, so that nuclear power will not have increased its proportionate share of the energy business, and conventional fuels will still be providing about 75 percent of our needs—at greatly increased costs, it should be pointed out.

The setback in nuclear output caught the power utilities without enough coal, and in 1969 the United States burned something like 8 million tons more of that fuel than it mined. In August of 1971 alone there were sixteen power brownouts and blackouts along the Atlantic coast. Predictions are for more of the same, and already environmentalists are being put off by power producers using fuels that fail to meet standards for pollution. As Glen Seaborg, former chairman of the AEC put it, the cries of environmentalists over pollution by power plants will be as a murmur compared with the enraged screams of a populace deprived of heat when it is cold, and air-conditioning when it is hot.

The idea remains widespread that nuclear power plants will solve at first running all our pollution problems. This is a good way from the facts in the matter. Aside from the dangerous radiation

FIG. 5. Artist's drawing of fast breeder reactor. The design and construction of this nuclear power producer has been beset by problems, and even if successful is still decades away.

produced by reactors, nuclear plants are notoriously poor converters of fuel to power, and produce even more "thermal pollution" than conventional steam-electric plants. Indeed, it is said that one of the heritages of nuclear power will be a lot of fried oysters in Chesapeake Bay.

Nevertheless, construction of nuclear electric plants continues in the hope that most electric power will be produced by such sources by the year 2000. Advertising for nuclear power includes exhortations for children to "go play in the nuclear power park," and the information that there is a nuclear plant practically next door to the California White House.

Even if radiation danger proves to be less than that forecast by opponents of this power source, nuclear electricity may not necessarily become a major contributor to our power needs. Like the fossil fuels of coal, oil, and gas, nuclear ores are finitely limited, too, and will ultimately be consumed. They are irreplaceable. Presently work is going on toward the "breeder" reactor, a more efficient converter of nuclear fuel to power in that it produces a by-product that is itself a fuel. The breeder reactor has not yet been successfully operated, however, and should it prove not feasible, projections of the share that nuclear energy could contribute would have to be revised sharply downward. This would of course delight many of the ecologists who claim that nuclear radiation is too big a risk to take for the rewards it offers in power production.

Even assuming that nuclear fuels are safe, then, they are by no means the be all and the end all of our power needs. Like fossil fuels they are limited, and within decades they may be gone. The power of the atom released in fusion, the multi-million-degree heat generated by the sun's conversion of hydrogen to helium with a by-product of tremendous energy, has tantalized scientists and engineers for more than twenty years, with billions of dollars spent by our country, Russia, and others. Theoretically, success in this endeavor would solve our power problems for the foreseeable

FIG. 6. Radiation monitor checks on radioactivity during burial of worn-out nuclear equipment.

future. To date, however, no fusion accelerator has operated successfully, and despite glowingly optimistic reports at intervals, there is no assurance that one ever will. The problem has been likened to harnessing a blowtorch with a rubber band; thus far the rubber bands melt through in a fraction of a second.

The Mathematics of Futility

To further dampen the hopes of those who imagine untold riches of energy waiting exploitation in the earth, there is a much more ominous warning than that of the economics of fuel production. For example, Claude Summers, writing in the September 1971 issue of *Scientific American*, approaches the consumption of fuels in another manner. Assuming that we have used 0.1 percent (one-thousandth of the total) of our fuel, and that we are doubling our consumption every ten years (which we are), Summers points out that we will consume *all* the fuel—no matter how much it was to start with—in just one century!

If we assume, much more optimistically, that we have consumed only 0.01 percent (one ten-thousandth of the total) and still have 99.99 percent of it remaining, we can survive somewhat longer. It works out to 133 years. Even estimating that we have used only one-millionth of the energy we inherited with the earth, we will still use it *all* in 266 years if we continue to double our consumption of fuels every ten years.

Here is no open question, no room for guessing, hedging, or arguing—only the deceptively simple mathematics of doubling consumption each ten years. No matter how much money is in the bank, or how much fuel is in the earth, it won't last long as consumption increases.

Ultimate Trap: The "Thermal Ceiling"

Point A

In producing power, engines produce heat as a by-product. This waste heat represents inefficiency in the conversion of fuel to power, but there is no way to avoid such loss in the thermodynamic cycle. Wasteful as such a loss is, there is an associated environmental danger that may in the long run be more of a threat to the environment than smog or the other forms of material pollution plaguing air and water and earth. This new danger has been called the "thermal energy ceiling."

Much has been said about city dwellers increasing the environmental temperature in hot weather by running refrigeration units. This increase in outside air temperature makes them run the units longer to keep cool, thus producing even more heat. The thermal energy ceiling represents the ultimate of such a losing battle against heat. It has long been feared that by burning hydrocarbons man is increasing the amount of carbon dioxide in the air and also the temperature of the atmosphere.

Within the past century, carbon dioxide concentration in the atmosphere has increased from 290 parts per million to 320 parts per million. By the year 2000 it is estimated that it may have climbed to between 375 and 400 parts per million, an increase of

FIG. 7. Nuclear power plant designed to eliminate thermal pollution of river. However, it will add heat to the atmosphere.

about one-third over what it was a century ago. Martin Wolf, of the Institute for Direct Energy Conversion at the University of Pennsylvania, suggests that by the year 2020 we may have increased the ambient air temperature over the United States by as much as 2.5 degrees F. While it would be possible for individuals to tolerate a temperature increase of ten times that (as residents of hot desert areas already do), the ecological balance of Earth is delicate, and some scientists feel that a few degrees could upset things drastically.

Claude Summers, quoted earlier on energy consumption, agrees that even more crucial than the consumption of fuel is the thermal barrier. He suggests that by the year 2000 most power plants will be located several miles off shore so that waste heat can be dumped (for a while at least!) into the sea. Summers points out that presently in the United States we are dumping waste heat into the environment at the rate of about 0.017 watt per square foot. Using the doubling factor approach for the production of heat as he does for the consumption of energy (the two go hand in hand) he points out that in about a century from now we will be dumping 17 watts into the atmosphere. The special magic in that number lies in the fact that it approximates the amount of heat the Earth receives from the sun!

Other Ways to Go

Commendable research work is being done with a conversion method called magneto-hydrodynamics (MHD). In this still experimental technique a plasma, or charged gas, is passed through a magnetic field to produce electric power. Potentially MHD offers a higher efficiency than conventional conversion methods. More than $50 million has been spent on the research and development of fuel cells that convert liquid fuels directly into electricity. However, the fuel cell has been known since 1839, and it has not yet reached the practical stage. Many power experts have felt that it never will because of the delicate and demanding nature of the chemical reaction that produces electricity.

More millions are being spent in developing techniques for producing gas from coal. "Synthetic" petroleum also seems destined to come. Chemists know how to make oil from just about anything containing carbon, but the cost is high, and again this process often uses up materials that could be used for other purposes than fuel.

Grandiose schemes have been proposed for such things as "geomagnetic power generators," in which a belt of satellites would tap the electricity potential in the geomagnetic force surrounding the earth. There are also dreams of harnessing "antimatter" as the end to all our needs. Perhaps some day these far-out power plants will be in use, but they seem centuries off at the moment.

Today the cry, not just of ecologists but also of many formerly apathetic persons, is for clean power. Instead of burning dirty fuel or releasing potentially dangerous nuclear radiation, why not take advantage of wind and water, and also of geothermal energy, which will not mess up the environment?

Hydroelectric power can undoubtedly contribute more than it does at present, but its total potential is limited, and it will be a constantly diminishing portion of the total. The same unfortunately seems true of wind power and tidal power, both of which are fixed in amount. There is much current interest in further exploiting geothermal energy. Again, this is an old idea, and has been done in Italy, New Zealand, Australia, and even Iceland. While it is a commendable use of a low-pollution form of energy, geothermal power suffers from the same shortcoming as hydropower—it is in relatively short supply. Less appreciated is the fact that it is a thermal polluter, too, in that it releases terrestrial heat faster than nature would.

Solar Energy: The Clean Alternative

Ironically, the model for the fusion power plant, the sun, is a source of practically unlimited energy, most of which is wasted but that nevertheless provides us with millions of kilowatts of power, keeps us warm, and grows all our food. To top it off, solar energy is *safe*, pollution-free energy on and in which living things have thrived since they first appeared on Earth.

Every day the sun showers Earth with several thousand times as much energy as we use. Even the small amount that strikes our roof is many times as much as all of the energy that comes in through electric wires. With the sun straight overhead, a single acre of land receives some four thousand horsepower, about equivalent to a large railroad locomotive. In less than three days the solar energy reaching Earth more than matches the estimated total of all the fossil fuels on Earth!

This book is about solar energy, the deferment of whose exploitation may someday be known as a national—or global—disgrace. Solar energy, unlike nuclear energy, has no "critical mass," and indeed no hazards at all except the possibility of sunburn, which is seldom fatal. Nor are there any waste products to be disposed of. It does not take a Ph.D. degree in physics to make use of solar energy in one's back yard.

Most people don't know the potential of solar energy; many of those who do simply shrug as they reach for the switch on the air-conditioner, or mash on the gas pedal to arouse the several hundred horses under the hood of the family car. Some ridicule the notion of harnessing sunbeams as on a par with recapturing the waste energy of golfers in America (millions of horsepower-hours on a good weekend).

We have been able to afford spending $20 billion on the moon in a decade; the tentative budget for exploration of Mars is about $50 billion. We have a great pool of scientifically and technically trained talent ready and eager for a new challenge as the jobs they prepared for vanish. Yet in all the doom-crying of ecologists scarcely a word is mentioned about solar energy, the beneficent radiation that makes life possible, grows all our food and timber, provides wind and water power that is nonpolluting, and does its best to clean up the pollution our machines contribute. The sun is ready and able to do a lot more for mankind. But not without man's help.

While the sunne shineth—make hay.

John Heywood (1546)

2 A LOOK AT THE SUN

Our world is not noted for its conservation of anything, as its conspicuous consumption of power attests. It is thus fortunate that the sun bombards us each day with many times as much energy as we are able to use in twenty-four hours. It has been doing this since the dawn of our solar system, and will likely continue its generous performance for billions of years in the future. For the maker of hay, at least, this beneficence is a never-ending blessing. The rest of us might well consider this primary source of energy that we call the sun, so that we can better understand its tremendous potential.

Our sun is actually a star, an orange dwarf at that, although the term is relative, since the sun's mass is more than 330,000 times that of Earth, and is about 2 thousand trillion trillion tons. If the Earth weighed but an ounce the sun would tip the balance at more than 10 tons. We are principally aware of the sun because it shines. This sunshine reaches across 93 millions of miles of space in a bit over eight minutes because it is electromagnetic radiation and travels with the speed of light. Most of this radiation occupies the spectrum from about .25 micron to 3.0 microns (millionths of a meter) in wavelength. About 9 percent of it lies in the extremely short, and invisible, ultraviolet region; about 40 percent is the visible light we use in seeing; and the remaining 51 percent is infrared, or long, waves.

Infrared rays account for the sun's heat. We learn from experience that the sun will burn our skin if we are careless at the beach,

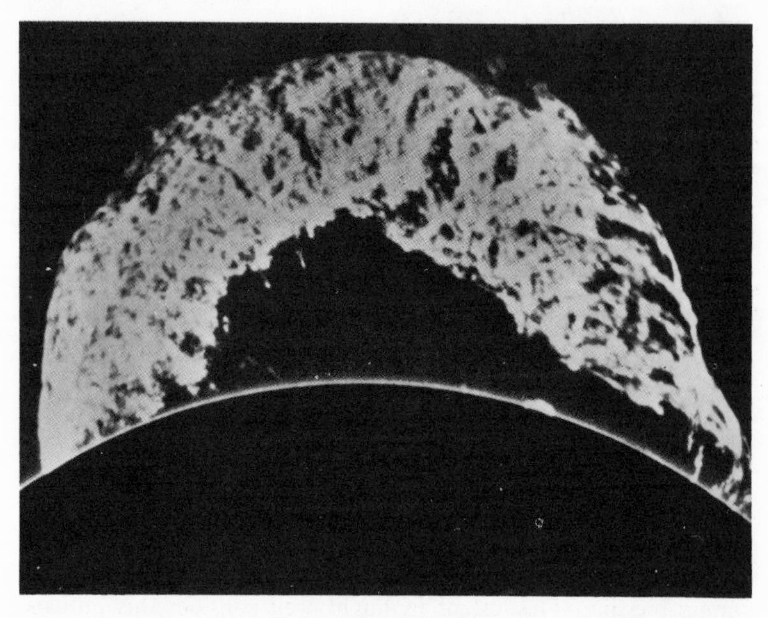

FIG. 8. Spectacular "prominence" erupting from the sun.

or our fingers if we are clumsy in using a burning glass. When we remember that this heat has come to us over such a magnificent distance, and when we realize that Earth and all eight of its sister planets together intercept only about 1/120 millionth of the total radiation, we realize that the sun is indeed a great ball of fire. How hot, we have only recently learned.

Some two hundred years ago the astronomer Herschel suggested that, although the surface of the sun was obviously quite hot, inhabitants might live cosily inside it, kept comfortably warm by the "furnace" above them. We have since satisfied ourselves that this naïve concept was gravely in error and that the inside of the sun is even hotter than its outside.

The fusion of hydrogen into helium, whereby the sun is continuously converting its substance into radiant energy, takes place at an estimated 30 million degrees F. This is far too hot to be called burning, or any other reaction we are familiar with. Even the sun's corona is believed to reach about 2 million degrees. Thus the surface, at 10,000 degrees F. or so, is relatively cool, although man can produce such a temperature only with sophisticated techniques like electrical plasma generation, the shock tube, and so on.

Prodigal Power Plant

As a result of the fusion inferno taking place in the sun (which consumes about 140 trillion tons of its mass each year), an appreciable amount of energy strikes the Earth. The "solar constant,"

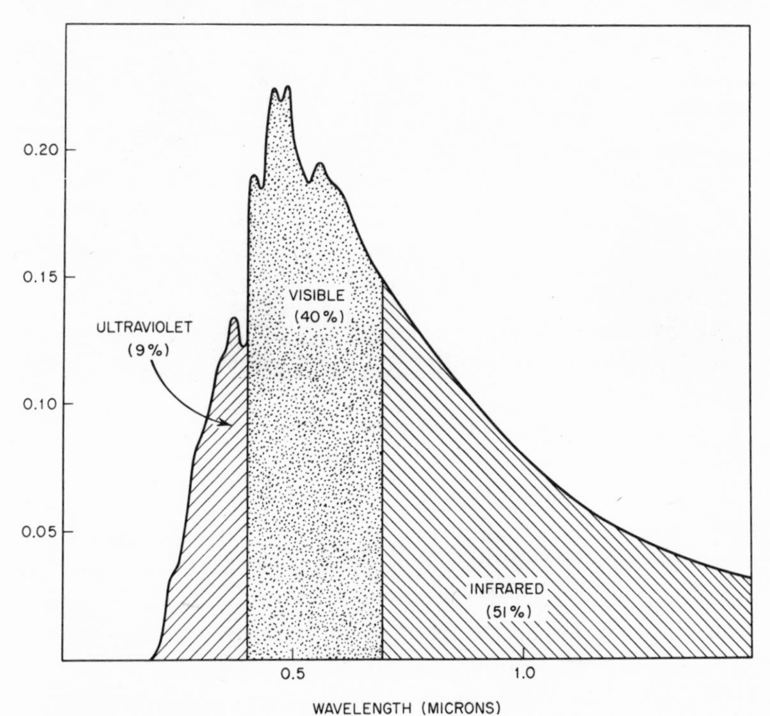

FIG. 9. Graph showing the amounts of sunlight in various parts of the spectrum.

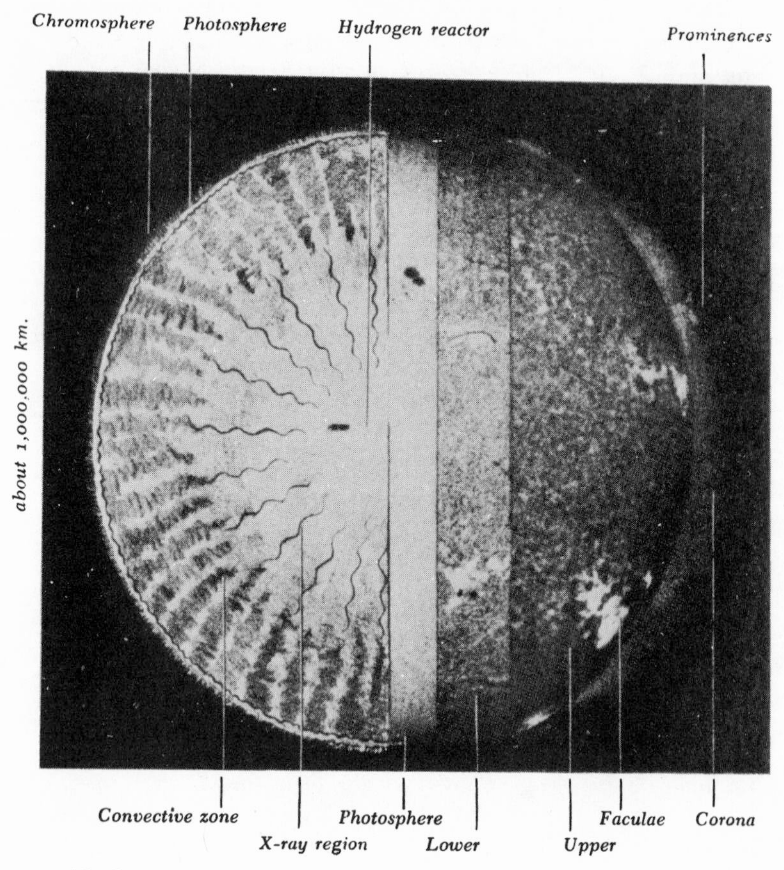

FIG. 10. Composition of the sun.

the average amount of sun energy reaching the Earth's atmosphere, is called the "langley," and amounts to two calories per square centimeter per minute. Translated into terms we can more readily visualize, a square yard of area exposed to direct sunlight continuously receives radiation equivalent to nearly two horsepower from the sun!

This value is a maximum, and is only approached atop high

THE SUN AND SOLAR ENERGY

Diameter of the sun	864,000 miles
Mass of the sun	2,200 trillion trillion metric tons (approximately 330,000 times that of Earth)
Temperature of sun's surface	10,000 degrees F.
Temperature of sun's center (est.)	30 million degrees F.
Solar mass consumed per second	4 million tons of hydrogen (converted to helium)
Total radiation constantly released by the sun	380 billion trillion kilowatts
Solar radiation reaching outer atmosphere of Earth	173 trillion kilowatts (less than one-thousandth of a millionth of total)
Solar radiation reaching Earth's surface	85 trillion kilowatts
Solar energy received in the United States annually	9,000 trillion kilowatt-hours (equivalent to 1,150 billion tons of coal)

mountains or in very clear air. Clouds, haze, dust, and smog cut the amount of energy received, and London and Los Angeles are not likely candidates for solar energy installations. However, a quarter section of land is showered with energy equal to that delivered by an oil well producing twenty-five hundred barrels of crude oil a day.

The mean value of the solar constant is 1.395 kilowatts per square meter. The total radiation continuously intercepted by the Earth (1.275×10^{14} square meters) is 1.73×10^{17} watts. This is about 173 trillion kilowatts, or 232 trillion horsepower.

About 30 percent of the solar energy striking Earth's atmosphere is immediately bounced back into space as short-wave radiation. About 47 percent is absorbed by the atmosphere, the land, and the oceans to contribute to the temperature of the

AVERAGE SOLAR ENERGY RECEIVED AT SELECTED CITIES

	Million kilowatt-hours per acre per year
Boston, Massachusetts	5.2
Cleveland, Ohio	6.1
El Paso, Texas	9.5
Fresno, California	7.8
La Jolla, California	7.0
Lincoln, Nebraska	6.3
Miami, Florida	7.0
New York City	4.9
Salt Lake City, Utah	6.7
Seattle, Washington	5.4
Washington, D.C.	5.8

environment. About 23 percent is used in the evaporation, convection, and precipitation processes of the hydrologic cycle. A small fraction of 1 percent powers the movements of air and the circulation of oceans, and is dissipated into heat by friction. An even smaller amount, about 40 billion kilowatts, is converted into plant energy in the chlorophyll of green leaves. It is this tiny fraction of solar energy that has produced all the fossil fuels we inherited with Earth. Of course, it also produces our food, timber, and other vegetation.

The sunlight intercepted at the outer atmosphere of Earth has an energy content estimated at 5,300 Q per year. This is the equivalent of more than 200 *trillion* short tons of bituminous coal. Every hour the sun showers Earth with about 0.6 Q. Thus, in a day the input is more than 14 Q, and in less than three days we have received as much as some estimates of total fossil fuels remaining on Earth. In forty days we receive enough solar energy to last a century—if we could make use of it.

To put these figures for solar energy into better perspective, we

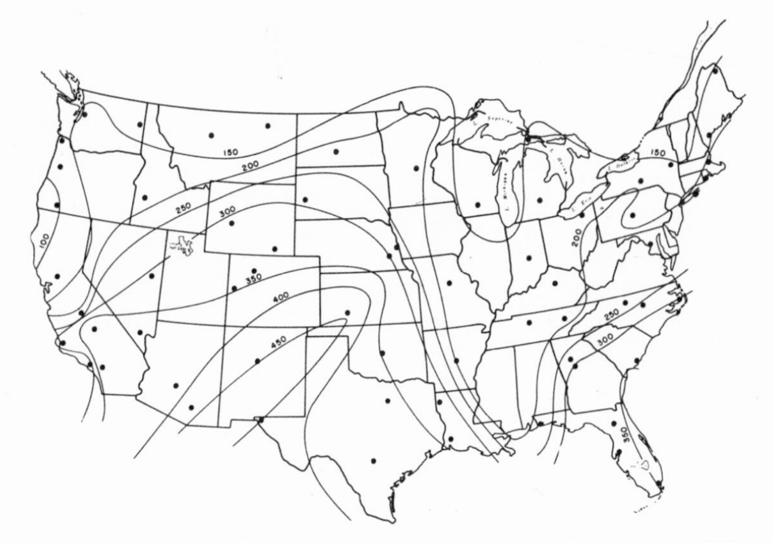

FIG. 11. Typical amounts of solar radiation received in different parts of the United States.

can compare them with our present use of power from other sources. In 1972 the United States used about 30 trillion horsepower-hours of energy (world consumption was about three times that or roughly 90 trillion horsepower-hours). It is indeed fortunate that we do not have to stable that amount of horseflesh. Yet, while the world was burning up this mountain of energy, the sun was sending us more than 1.5 *million* trillion horsepower-hours.

We presently use up to twenty times as much energy for our machines as we do to feed ourselves. Projections for the future estimate as much as one hundred times the energy requirements for machines, a fivefold increase in the use of mechanical energy as compared with food. Today solar energy provides us with little more than food. Potentially it might provide up to one-third of our energy needs for other things.

Nuclear power, by its nature, falls in the domain of government and requires the spending of huge sums for its development. However, though the moon seems in danger, nobody owns the

sun; its energy falls in all our back yards. Our material standard of living depends on the energy available to us, and solar energy falls with no regard for race, color, or creed. It obeys only the laws of physics and of geography, and thus we have a wide "solar belt" between latitudes 40 degrees north and 40 degrees south within which there are large amounts of available power. As good fortune would have it, most of us live in this solar belt, and many of the "poor" parts of the world are blessed with greater solar riches than their presently more wealthy brothers. Nature has provided for us far beyond our dreams, in spite of the easy lives we have led on the fuel reserves we inherited with the Earth.

An acre of ground can be expected to provide crops of about three tons dry weight per year. Dr. Farrington Daniels of the University of Wisconsin points out that it is *theoretically* possible to produce this same amount—three tons—in a single *day* on the same acre of land! Obviously such a goal is impossible of achievement, but it does give promise of the improvement that may be made. Work with algae called *chlorella* has indicated that perhaps twenty tons a year per acre can be grown, and if the algae can be persuaded to grow in warm or hot water, even higher yields are possible. One need for such developments ought not to be missed, however: even today's efficient, mechanized farming requires about a calorie of mechanical energy to produce a calorie of food, and this is an added drain on our fuel supply.

Solar energy is available in whopping quantities, it is free, and it requires no transportation or maintenance. But to the man who now buys electric power conveniently for as little as a penny a kilowatt-hour, solar energy remains only an interesting scientific curiosity.

Light Harness for the Sun

A logical question at this point is, Why are we not making use of this incredible bonanza in the form of solar energy? The answer, of course, is that we *are* using it, and have been from the begin-

FIG. 12. Solar energy used indirectly as wind power to pump water for sheep.

ning. All our energy—except nuclear—comes originally from the sun. A solar-powered radio draws on the sun directly, but a gasoline-fueled automobile also uses solar energy—stored solar energy, in which the sunshine of ages ago was trapped in the earth until reclaimed by oil drillers.

The hydroelectric plant is an example of solar power, though it is interesting to note that the energy in sunshine falling on the surface of Lake Mead is five times the output of the generators at Hoover Dam! The windmill depends on the sun, and even the tides may be considered a form of solar—and lunar—energy, since they depend on the pull of those celestial bodies. There are a few tidal power schemes in operation today, but their total output is tiny compared with conventional power plants.

Food and lumber come to us courtesy of the sun; so does rainfall, and so do the fish we get from the sea. We were born of the sun and depend on it for sustenance. But outside this more or less passive acceptance of the sun's bounty, we have done little to exploit its energy potential. The ancient Egyptians believed that the sun-god, Ra, journeyed during the night through an underground and hellish river so he might rise again next morning in the east. There are still people in some lands who think that the sun steals back across the dark night sky. But we, in the sophistication of our laughter, are perhaps more ignorant than they. While the savage at least worships the sun, we know little of it, care not much, and do even less about it.

The sun offers a tremendous potential in energy, not just for the few but for every human being on Earth. The power is there for the taking. This taking, unfortunately, is not easy for men spoiled by the luxury of stored energy—not as easy as digging gashes in the ground and mining coal, or drilling holes into handy subterranean storage tanks of gas and oil. But the harnessing of sunlight should be a far easier task than that of unleashing nuclear energy on which we have spent many billions of dollars and which remains a tricky and treacherous source of power.

The Trouble with Solar Energy

Of course, it is not just man's perverseness that keeps us from making more and better use of solar energy. There are some very difficult problems, and not all of them are psychological. Some are technical and some are economic, and these various barriers make it difficult to change the status quo of energy conversion.

Sunshine is plentiful, but its energy is diffuse, or at a low temperature. This energy is also variable, and requires a storage system to make it available whenever needed. There is another stumbling block, too—the strong hint of magic or chicanery in the notion that we can pull energy out of thin air and sunbeams.

Among the factors that make up the conspiracy against exploitation of solar energy, perhaps the strongest is that of inertia. The belief that "If it was good enough for Grandpa, it's good enough for me and the kids" sums up this phenomenon. So you can heat water with the sun, or cook a meal on a reflector stove. So what?

You could also probably cook with music if you worked long and hard enough at it, but what would be the point? Meantime you can still get a million Btus in a quarter's worth of kerosene, and it is so easy to strike a match or turn a valve or push a button.

Cheap conventional fuels are an obvious reason for not pushing solar energy applications. A camper can buy a gasoline stove for half what a folding solar reflector would cost, and it does little good to point out that he could also use the latter for an umbrella when it rains. The prospective buyer of a solar house-heating system is faced with a capital outlay much greater than that required for a gas or electric system, against the promise that his solar heater will be economical over a long period. Solar energy is free, to be sure. But the taking isn't.

Even though a square mile of sunshine seems to be equivalent to 3 million horsepower, engineers shudder at the thought of erecting a collector that size. The fact that 75 percent of the sunshine hits water is a consideration, too. It is easier to rock along on coal and oil, sure that Providence will not let us down. As luck would have it, Providence hasn't—up to now.

The fictional Russian philosopher, Kuzma Prutkov, decided that the moon is more useful than the sun, since it shines at night when light is needed; while the sun is of little use in daytime since it is light anyway! In such a fashion we, too, have dismissed the importance and potential of the sun. Astronomer Donald Menzel put his finger on the reason for our apathy; he likens the sun to a husband who is so dutiful and dependable that he is not appreciated. In fact, the sun's very regularity keeps us from even noticing it. The squeaky wheel gets the grease, and the sun is generally just too quiet.

The fruitful application of solar energy has been doomed to wait until the day of reckoning was nearer at hand, when man would actually scrape the bottom of the stockpile the sun had willed him. Had it not been for this stockpile he would have learned long ago to use direct solar energy. Later, had it not been for the apparent promise of the nuclear fuels, he might belatedly have begun to learn. But man, a natural being, behaves in a natural way. Like water obeying the law of gravity, he seeks the easiest course.

The Energy Gap

Fossil fuels are thought to have been produced over a period of about 600 million years. The process is probably continuing, and at about the same rate. In another million years, for example, more fossil fuel will be produced in an amount of about one six-hundredth of the total already formed. We are using it up much faster than the sun replaces it, and in the last century we have withdrawn nearly twenty times as much capital energy as we did in the previous seven centuries.

For those who look beyond themselves to the generations who will inherit the future we make for them, the picture is not particularly bright until we turn to solar energy, the primary source of all but nuclear power, and the *only* supply that is inexhaustible. Not even the most avid solar buff suggests switching to 100 percent solar power tomorrow, but we might well *begin* exploitation

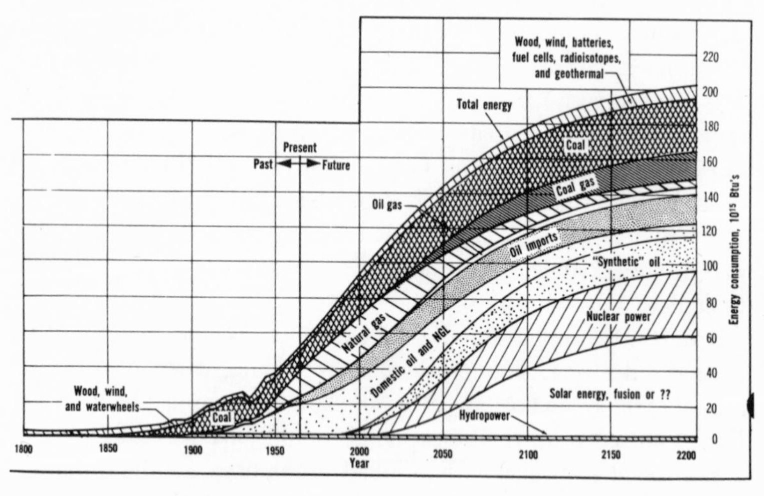

FIG. 13. Graph of past, present, and future use of energy sources.

of direct solar energy as a supplement to our fossil fuels, and to nuclear energy.

We have noted that in recent history a new source of energy has been developed every several decades. In 1870 coal began to displace wood, waterpower, and wind power. Some forty-five years later gas entered the fuel picture. Oil was discovered in 1859, and hydroelectric power became a factor in about 1890. And of course the most recent was nuclear energy, with the first such power plant completed in 1957.

Leon Gaucher, formerly on the research staff of Texaco, Inc., feels that this cyclic process must be repeated, and that by the year 2000 another major source of energy will be producing power for the world. This could be nuclear fusion, he says, but "it could most likely be solar energy, providing the research directed toward that end is mounted soon and pursued diligently."

As seen in the accompanying figure, prepared by Gaucher, both population and energy consumption are destined to continue their climb. Coal, natural gas, oil, and hydroelectric power are shown as the major contributors to power needs. Nuclear energy is not shown because at this point its contribution is minimal. Even a century from now, nuclear energy will only be matching coal in producing power (and gas made from coal will be providing additional amounts of fossil power). Natural gas will be declining, and oil will contribute about the same percentage of the total it now does. However, there is a large unshaded area amounting to about 20 percent of the total needs that must be filled with a new source of energy, unless we cut our demands for power. As the years go by, this 20 percent gap will increase to 30 percent by the twenty-second century.

Solar Supplement

Surprisingly, the portion of total energy represented by such sources as wood fuel, wind power, and geothermal energy is shown increasing markedly on Gaucher's chart, an indication of the growing shortages of conventional fuels. However, it cannot be hoped that these "natural" sources can fill in the energy gap already on the horizon.

The idea of reverting to wood for producing power has been advanced, and investigated thoroughly. It seems an attractive proposition, in that man would be living off the income of solar energy. However, hard figures show that if *all* our arable land were used for wood, or even for advanced and efficient algae culture, the fuel produced would provide only 10 percent of the requirement of the world's people. We would also be homeless—since most houses are built of wood—and would starve to death in the bargain.

Wind power is the result of solar energy, of course. It has been estimated by the World Meteorological Organization that there is a wind-power potential at favorable sites around the world of about 20 million megawatts. This is about sixty times the present generating capacity of electrical power plants in the United States.

For all the attractiveness of wind power there are several technical problems that seem to prevent the implementation of this solution to our power needs. For example, the largest feasible wind turbines seem to be only a few megawatts in capacity, and an estimated 3 million of them would be required to take care of U.S. needs. In recent years Russia reportedly planned construction of some 600,000 windmills of two-megawatt output for use on the plains where moderate but steady winds were available. However, there is no indication that this phase of one ten-year plan was ever accomplished. Unlike other solar-energy conversion methods, wind power has been exploited for many centuries; it has been tried and apparently found wanting.

At present only about 8.5 percent of hydropower is developed. The maximum potential is estimated at about 3 billion kilowatts, about as much as is presently used by industry worldwide. Most of the undeveloped waterpower is in Africa, South America, and Southeast Asia, all of which have economic problems in such development.

Geothermal power, presently used to an appreciable extent in some places, and boasting a 370-megawatt plant at Lardarello, Italy, has a small potential. Unlike tidal power it is limited, and estimates are that if it were all exploited within a fifty-year period it would produce about 60 million kilowatts of power, roughly the same as tidal power. Of course, at the end of that time there would be no geothermal energy remaining in the earth.

Tidal power has a potential estimated at about 64 million kilowatts, only about 2 percent of the hydropower potential.

Only solar energy, in more direct form than wind power and wood fuel, seems to offer the source that can make up the energy gap already plaguing the developed nations. (The underdeveloped lands have even harsher problems, which has led in some of them to fining citizens for burning cow dung as fuel because it is desperately needed for fertilizer!) Gaucher is convinced that solar energy must be used perhaps within thirty to fifty years—not just as a source of power but also to produce the fuels and chemicals needed in our technological society, and which are presently being burned up to produce power.

Although the first nuclear power plant was started up more than fifteen years ago, nuclear electricity has yet to make a dent in the total needs of the nation. This is evidence of the lead time necessary to phase in a new energy source. Solar energy will be no exception—its development should begin now if it is to be counted on in our future.

. . . there is no new thing under the sun.

Eccles. I:9

3 THE SOLAR PIONEERS

The sun naturally figures in man's earliest attempts to probe the world around him. So obviously was man dependent on the sun, early religion and culture were built around that celestial body. Ra the sun-god and his Heliopolis are well known in ancient Egyptian lore. Greek and Roman deities included Apollo and Phaeton, with the sun itself a fiery chariot driven across the face of the sky. Druids and Aztecs, oceans removed, worshiped the sun. Around the world and down through the ages, man has lifted his eyes to the sun. First with awe and fear that prompted human sacrifices to the god in the sky, and more recently with gratitude and even curiosity.

The sun has been both blessing and challenge as far back as Icarus, who dared fly too near it and perished. Somewhere along the way myth and legend begin to give way to fact. Surely temple fires were lighted, and doors swung open, by sunlight, but something more impressive happened in 212 B.C., the year in which Archimedes reputedly set fire to an attacking Roman fleet. "With burning glasses," says Galen, "he fired the ships of the enemies off Syracuse." Twelfth-century author Johannes Tzetzes wrote a detailed account of the fleet burning with the new wonder weapon. "Archimedes set fire to Marcellus' navy by means of a burning glass composed of small square mirrors moving every way upon hinges which when placed in the sun's direct rays directed them upon the Roman fleet so as to reduce it to ashes at the distance of a bowshot."

FIG. 14. King Tutankhamen and Queen Ankhesenpaaten, showered with blessings from the sun-god, Ra.

The Coming of Solar Science

Whether or not Archimedes did set fire to the sails of attacking ships remained one of the controversies of history, and there were many who scoffed, calling it confusion of legend with fact. How-

FIG. 15. Reconstruction of Archimedes' storied burning of an enemy fleet with a solar furnace.

ever, the place of the sun in man's thoughts changed. In 1350 B.C. the sun-god Ra had shone protectively over King Tutankhamen and his queen in Egypt, but by A.D. 1600 science had begun to look questioningly beyond superstition and magic.

Coincidentally with Galileo's scientific pioneering with his solar telescope, another man was harnessing the energy of the sun for mechanical use—perhaps the first such feat in history that was more than a stunt. In France, Salomon de Caus put the sun to work heating air in his solar "engine," used to pump water. De Caus was more than two hundred years ahead of the next inventor of a solar engine, but the gap was filled with simpler devices.

Descartes had suggested that either Archimedes' mirrors were huge or else the whole story was a fable, but other scientists did not agree. Among these was one named Kircher, who claimed in the seventeenth century that his array of five mirrors actually burned material at a distance. A century or so later the Frenchman Georges Buffon proved once and for all that, whether or not

ISTORIA
E DIMOSTRAZIONI
INTORNO ALLE MACCHIE SOLARI
E LORO ACCIDENTI
COMPRESE IN TRE LETTERE SCRITTE
ALL' ILLVSTRISSIMO SIGNOR
MARCO VELSERI LINCEO
DVVMVIRO D'AVGVSTA
CONSIGLIERO DI SVA MAESTA CESAREA
DAL SIGNOR
GALILEO GALILEI LINCEO
Nobil Fiorentino, Filosofo, e Matematico Primario del Serenis.
D. COSIMO II. GRAN DVCA DI TOSCANA.

IN ROMA, Appresso Giacomo Mascardi. MDCXIII.

CON LICENZA DE' SVPERIORI.

FIG. 16. Galileo's report of his telescopic observations of the sun.

Archimedes actually did burn sails at a "bowshot," the deed could have been done. In 1747 he put together an array of 168 small flat mirrors and succeeded in burning wood 200 feet away—surely a good bowshot. And lest there still be skeptics, he went on to melt lead at 130 feet and silver at 60 feet! Buffon concluded that Archimedes could well have burned cloth sails at the distance claimed. Now it was time to make practical use of this phenomenon, and in the same century as Buffon's feat an optician named Villette, of Lyons, France, built polished-iron solar furnaces that could smelt iron, copper, and other metals. His furnaces were used not only in France but in Persia and Denmark as well.

In 1695 two Italians named Targioni and Averoni used a large burning glass to decompose a diamond, undoubtedly an experiment that taxed their budget. Other researchers using burning lenses at the time were Parker in England and Tschirnhaus in Germany. The duke of Orleans bought two of Tschirnhaus' lenses, measuring thirty-three inches in diameter, and these were used by French scientists. Among the work they did, unfortunately, was "proving" the validity of the "phlogiston" theory then current that heat was a material substance.

In the late eighteenth century the French scientist Antoine Lavoisier finally undid the erroneous work of his predecessors and disproved the faulty phlogiston theory, using powerful lenses to focus sunlight and burn samples. In these experiments he clearly showed that there was no weight lost or gained, and that the "element" phlogiston was therefore not real. Lavoisier used curved glass discs fastened together at their rims, with wine filling the space between. The vintages used are not identified, but we do know that Lavoisier was furnished glass by the great Saint Gobain glassworks. (This same firm provides the mirrors used in France's Mont Louis solar furnace, and the American–Saint Gobain Corporation has worked with Dr. George Löf in this country on the glass collectors for his solar-heated home.) Using a fifty-two-inch lens plus a secondary eight-inch lens, Lavoisier achieved a temperature close to 1,750 degrees C. and almost succeeded in melting platinum. This was far beyond any temperature attained by man up to that time.

Lavoisier also carried the science of the solar furnace forward by heating samples in a vacuum and in controlled atmospheres, using quartz containers. He also noted that "the fire of ordinary furnaces seems less pure than that of the sun," a very important considera-

FIG. 17. M. Lavoisier with his two-lens solar furnace.

tion, as later researchers have brought out. It was Lavoisier's misfortune to live during the Revolution, and his fate to die on the guillotine because "the Republic has no need for scientists." With the lopping off of this pioneer's head, work with solar furnaces halted, and it was more than a hundred years before men again achieved elevated temperatures using the sun's rays.

Power from the Sun

In 1816 the Scottish clergyman, Robert Stirling, and his brother, James, built a two-piston air engine. Although it was not designed to be operated by the sun, the Stirling cycle was proved ideally suited for such use. Swedish inventor John Ericsson introduced his hot-air engine in 1826, and used three-hundred-horsepower versions of it to power the paddlewheeler *Ericsson*. Later the engine was to be adapted for solar operation.

August Mouchot, with the backing of Napoleon III, developed in the period from 1866 to 1872 a steam engine powered by the sun. This effort was exhibited in Tours and later tested in Algeria for pumping water. Six years later Mouchot produced a larger version with an improved boiler arrangement. It was tested for six months in a water distillation plant, but the government decided Mouchot's engines could not be made with an economy "sufficient to the demands of commerce."

John Ericsson had meantime come to the United States, and was famous during the Civil War for his *Monitor*. Ericsson had built many successful hot-air engines, and it was not surprising that he turned his inventive mind to using "the big fire hot enough to work engines at a distance of 90 million miles." In 1868 his work in this direction earned him an honorary Ph.D. from the University of Lund in his native land.

Ericsson claimed his solar engine, "operated by the intervention of atmospheric air," was the first of its kind, and also that an earlier solar steam engine held the same distinction. These solar engines may be seen in the American-Swedish Historical Society in Philadelphia. By 1875 Ericsson had built eight different models of his solar engines, but although he claimed high efficiency for the designs none of them was practical. In 1883 he made one last try, building what was the second largest solar engine of that day. Measuring eleven by sixteen feet, the rectangular parabolic collector drove a piston with a six-inch bore and eight-inch stroke. Designed to work with steam or air, the engine was connected to a five-inch pump, and Ericsson claimed it delivered one horsepower for each hundred feet of collector, or nearly two horsepower with the collector he used.

Discouraged with the immediate economic prospects of solar energy as a competitive fuel, even though he felt "the field waiting it is almost beyond computation," Ericsson shrewdly converted his solar engines to run on coal or gas and dried his eyes on the way to the bank. More than fifty thousand Ericsson hot-air engines were sold all over the world, repaying the inventor several times for the expense of his solar research.

In 1876, shortly before Ericsson built his big solar engine, a man named Adams put together an amazing power plant. Built in

THE UTILIZATION OF SOLAR HEAT FOR THE ELEVATION OF WATER.

FIG. 18. An early attempt to use solar energy to pump water. Heat collector panels are numbered, and provide heat source to pump water shown at G.

Bombay, this 2.5-horsepower steam engine was heated by a huge hemisphere formed of ten- by seventeen-inch mirrors, a collector measuring forty feet from edge to edge! This solar giant was used as a pump. And in France about 1880 Abel Pifre built the first solar engine used in a commercial venture. His steam engine, fired by a hundred-square-foot parabolic collector, generated two-thirds horsepower, and Pifre put it to work running a printing press. The paper, understandably, was called *Le Journal Soleil* ("The Sunshine Journal").

Shortly after this achievement another Frenchman, Charles Albert Tellier, built a solar engine that was unique in more ways than one. Instead of using a focusing collector, as had all his predecessors except perhaps De Caus, Tellier used a "flat-plate" type. Its 215 square feet of area drove an engine with ammonia as a working medium instead of steam, air, or water. An illustration in the *Scientific American* in 1885 proposed a system in which the flat-plate collector served also as the roof of a factory building, an intelligent attempt to cut costs of the installation.

Time and again men proved that there was power in sunshine and put it to work. But all were ahead of their time, and the ideas did not take hold. As Ericsson sadly lamented, although solar energy was free, capital investment was so high for the necessary

FIG. 19. Dr. Charles Greeley Abbot, father of solar energy research in the United States, and one of his solar stoves.

collector and associated equipment that it cost much more to run a solar engine than a conventional type.

Using Solar Heat

The sun is an ancient cooking device; foods have been sun-dried for ages. But the development of the solar cooker dates back only some hundred and fifty years. Adams in Bombay, Herschel in England, de Saussure in Switzerland, and Mouchot in France pioneered the field, building simple "ovens" with glass lids to let in the sunlight and trap heat inside. Mouchot demonstrated his cooker at the World Exhibition in Paris in 1878, cooking a pound of beef in a little over twenty minutes. Samuel Pierpont Langley, whose solar research is honored in the "langley" as a unit of solar radiation, built such a cooker and demonstrated its use atop Mount Whitney in California. Dr. Charles Greeley Abbot, the father of American solar energy research, built a more sophisticated oven in 1916 and tested it for several years on Mount Wilson.

Through the ages man has exploited one manifestation of solar energy—the evaporation of water by the sun. The ancients harvested salt, and the practice is carried out today in much the same manner except for some refinement in the way the product is gathered and handled. There is, of course, a by-product to the production of salt—fresh water. In some areas this commodity is in critical supply, and in 1871 a huge solar still was built by an American named Charles Wilson in Las Salinas, Chile, to provide a mine with drinkable water. Another such still was operated at the Oficina Domeyko mine. The Las Salinas still provided six thousand gallons of water a day and operated successfully for forty years, yet today there is little to mark the location of this pioneering use of solar energy. Indeed, the method had been long forgotten when it was reinvented some time ago by French engineers who won a prize offered by their government for a means of providing fresh water from salt or brackish supplies.

The Engine Builders

In the United States, just before the turn of the century there was a reawakening of interest, and a number of fairly large solar engines made their appearance. One was installed at a Pasadena, California, ostrich farm to pump water. The thirty-three-foot apparatus is attributed to A. G. Eneas and a "band of Boston capitalists." Later these designers moved to a location where the solar engine had much appeal, the desert land of the Arizona Territory.

The Arizona Republican, a Phoenix newspaper, carried this description of the strange device in its issue of February 14, 1901:

> The unique feature of the solar motor is that it uses the heat of the sun to produce steam. As "no fuel" is cheaper than any fuel, the saving to be effected by this device is evident. When the solar rays have heated the water in the boiler so as to produce steam, the remainder of the process is the familiar operation of compound engine and centrifugal pump.
>
> The reflector somewhat resembles a huge umbrella, open and inverted at such an angle as to receive the full effect of the sun's rays on 1,788 little mirrors lining its inside surface. The boiler, which is thirteen feet and six inches long, is just where the handle of the umbrella ought to be. This boiler is the focal point where the reflection of the sun is concentrated. If you reach a long pole up to the boiler it instantly begins to smoke and in a few seconds is aflame. From the boiler a flexible metallic pipe runs to the engine house near at hand. The reflector is thirty-three and a half feet in diameter at the top and fifteen feet at the bottom. On the whole, its appearance is rather stately and graceful, and the glittering mirrors and shining boiler make it decidedly brilliant. . . .

Eneas, holder of a patent on the solar engine, and a number of backers operated it at several locations and apparently successfully pumped irrigation water. A price of twenty-five hundred to three thousand dollars was quoted for the solar devices, but—alas for the bright dreams of inventors, promoters, and editorial writers—the solar engines fizzled out for a variety of reasons. Mechanical

FIG. 20. Solar-powered pump built in the desert near Tempe, Arizona, by A. G. Eneas.

troubles and accidents plagued the installations, and windstorms finally destroyed the demonstration model.

Independent of the ostrich farm capitalists, two men with their own ideas for solar engines began experiments in 1902 with the flat-plate collector that Tellier had introduced in France. H. E. Willsie and John Boyle, Jr., built their early engines in Olney, Illinois, and Saint Louis, Missouri. In these engines heat was trapped in water flowing through shallow glass-covered basins, and used to drive water-ammonia engines. Such a collector was obviously cheaper to build than the complicated mirror types that followed the sun.

By 1905 Willsie and Boyle had headed for the land of sunshine and set up shop in Needles, California, a town then featuring sand and solar energy, and little else. Using about six hundred square feet of collector area, they operated a slide-valve engine that ran a water pump, a compressor, and two circulating pumps. In 1908 a second and larger plant was built at Needles. This one had one thousand feet of collector area and could store heat for later use. The solar heat generated a boiler pressure of 215 pounds and drove a sulphur-dioxide engine claimed to develop about fifteen horsepower. Estimated cost of construction for the plant was $164 per horsepower, still about four times the cost of a conventional steam plant. And even free fuel in the form of sunshine wasn't enough inducement to sell solar engines.

Fifty Kilowatts for Cairo

About 1907, in Tacony, Pennsylvania, an inventive engineer named Frank Shuman was beginning a career as the solar energy salesman who came closest to making the grade. A hardheaded realist, he favored the economical flat-plate collector idea used by Willsie and Boyle. His test engine had twelve hundred feet of collector area and produced a creditable 3.5 horsepower. Emboldened by success, he proposed not an engine but a huge solar steam plant that would cover four acres! The heat collector would be the entire tract of land itself, rolled level and formed into shallow troughs. A layer of asphalt waterproofed the trough, which

FIG. 21. Solar engine built by Frank Shuman in 1907. It used a flat-plate heat collector, shown to the left of the man.

was then filled with about three inches of water. Over this, Shuman proposed to pour a thin film of paraffin, apparently to help trap heat in the water. The whole area would then be covered with glass, leaving a six-inch air space between glass and water. This monumental project would store heat so that low-pressure turbines could be run continuously, and Shuman estimated an output of 1,000 horsepower! Optimistically he mentioned a cost of about forty dollars per horsepower, a figure to make the plant competitive with a coal-fired one at the outset. But in a day when men like Shuman were only wistfully beginning to dream of "a celluloid-like material having all the necessary properties of glass and being flexible and capable of manufacture in large sheets," such a plant was some time in the future.

Shuman found that he must use focusing collectors to increase efficiency, but he kept the design simpler than most. His next solar collector consisted of twenty-six long sheet-metal troughs, mirrored on their inner surfaces to reflect heat onto boilers. Instead of the twelve hundred feet of area his pilot model used, this new one boasted a whopping ten thousand feet plus. No toy, the solar power plant worked, and Shuman was confident it had a potential output of one hundred horsepower. It had cost him twenty thousand dollars, five times more per horsepower than the forty dollars he had hopefully predicted.

The Shuman plant did not deliver anything like one hundred horsepower in Tacony, near Philadelphia. There was air pollution even in those days, and smoke and cloud obscured the sky and severely hampered operation. But that was all right, because the solar plant had been designed to operate under clearer skies. Frank Shuman had formed the Sun Power Company in 1908. Now he obtained British backing for what was to be the Eastern Sun Power Company, Ltd., of London. He also had the technical assistance of Professor C. V. Boys, a noted physicist, and a go-ahead to build his one-hundred-horsepower plant in Egypt.

By 1912 the power plant was in operation at Meadi, a suburb of Cairo on the road to Helwan. Good as the design was that

Shuman had produced in Pennsylvania, the Egyptian model was better. Instead of the twenty-six flat reflectors used earlier, the new plant had only seven, spaced farther apart so as not to shadow each other. They were a parabolic shape to generate the highest temperature in the tubes producing steam at the focal point. These tubes were painted a special absorbent black.

As the sun moved across the clear Egyptian sky the reflectors, each 204 feet long, followed it automatically. This was accomplished with power from the solar engine, special gears, and thermostat sensing elements. The total collector area had been increased from the original ten thousand feet to more than thirteen thousand feet. Eagerly the builders put the first real solar power plant into operation, crossed their fingers—and watched it work! In one five-hour test run, Shuman's solar steam plant turned sunshine into work. The brake horsepower rating never dropped lower than 52.4 and once climbed to 63.0 horsepower. It was a proud moment for Shuman, even though the plant did not deliver its designed maximum. The economics of this plant, assuming it at 50 horsepower, have been worked out and show that it would have paid for itself in three years.

But good an engineer as he was, even Frank Shuman wasn't good enough to make solar energy stick. Eastern Sun Power Company, Ltd., had been formed to provide power for irrigation from the Nile, a job traditionally done by some 100,000 fellaheen, or Egyptian laborers. Caught in an advance skirmish of the battle of automation versus manual labor, plus economics and the confusion of World War I, Shuman's marvelous solar power plant was eventually abandoned and fell into disrepair. It must have crushed the inventor; surely it laid applied solar energy low for a long time.

So Near and Yet So Far Away

In the years that followed, scattered reports of solar applications made the news and kept the idea alive. In New Mexico an engineer named Harrington lighted a mine in a remote area with electricity produced by solar energy in a novel fashion. During the hours of sunlight a concentrator focused the sun's rays onto a boiler whose steam drove a pump to lift water to a storage tank some distance above ground. Water falling from the tank drove a dynamo to provide around-the-clock power output.

In 1919 a scientist doing research and development in an entirely different field made mention of the use of solar energy as an aid to that work. Dr. R. H. Goddard, who would one day be hailed belatedly as the father of modern rocketry, published a modest paper called "A Means for Reaching Extreme Altitudes." This means, of course, was with chemical rockets, but in his paper Goddard touched on the use of solar energy to aid in getting a rocket to the moon.

In 1924 Goddard filed for a patent for an "accumulator for radiant energy." By 1930 he had applied for a total of five patents pertaining to solar energy, and his solar power plant described in detailed drawings in 1929 looks very much like those being built today for use in space. Goddard wisely knew that the storage of

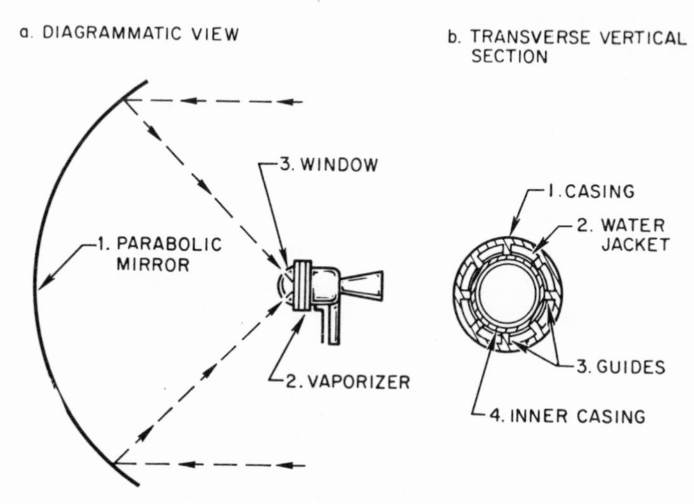

FIG. 22. Solar engine patented by rocket pioneer Robert Goddard.

heat was an important part of the solar engine, and his patents covered efficient means for doing this. Present-day "cavity flux-traps" resemble Goddard's early designs, and his use of a quartz window for permitting sunlight to enter directly into the chamber of his "vaporizer" anticipates the work of Battelle Memorial Institute scientists and their improved Stirling-cycle solar engine.

Claiming an efficiency of 50 percent for his solar power plant, Goddard foresaw tremendous possibilities for solar energy, particularly on farms. Moving up from the Earth he envisioned solar-powered aircraft—the fuel was weightless, he said—and even space travel accomplished with sun energy. An interesting proposal was a solar-powered dirigible. To get around the high weight of a large metal mirror, Goddard cleverly proposed "a thin layer of light fabric coated with highly polished metal foil." This was amazingly close to a description of the aluminized plastic presently being used for that very job.

As early as 1907 Goddard had hinted at solar energy for electrostatic space propulsion. Again, in 1911, he discussed the use of solar energy in space travel. It is a remarkable coincidence that the man who pioneered the rockets that are now blazing trails into space should also concern himself with solar energy. And it is unfortunate that he did not live to see the two disciplines converge that day in 1958 when the rocket Vanguard I blasted off with the first solar batteries aboard.

Dr. Abbot, by now the dean of solar science, continued his work patiently, and by 1936 had designed a solar steam engine that produced one-half horsepower. This engine was exhibited at the International Power Conference in Washington and caused a flurry of interest when it furnished the power for a nationwide radio broadcast.

From 1925 to 1935 the French scientist Georges Claude, better known for his development of acetylene gas and neon lighting, conducted experiments in tapping sea thermal energy. In this period Claude spent a fortune of his own money installing and testing both a submarine cold-water pipe from a power station off the coast of Cuba and a vertical pipe from the floating *Tunisie* power station. He was successful in operating a steam turbine on the temperature difference from top to bottom of the ocean, but performance was disappointing. The French government carried on his work to some extent, and in 1948 formed Energie des Mers for continued sea thermal research.

The solar battery, too, has a long history. In 1839 Antoine Becquerel found that sunlight caused a weak current in certain materials, in this case the electrodes of an electrolyte solution. Forty years later Adams and Day observed a similar effect in a solid material, selenium. The light meter using selenium cells, and "electric eye" applications were the main uses of photovoltaic conversion, with the converted electricity too feeble for anything except making a needle move on a scale, or activating a switch in a control operation.

In 1931 Dr. Bruno Lange demonstrated photovoltaic solar power at Kaiser Wilhelm Institute. His invention was a "sandwich" of copper oxide, silver selenide, and a third, secret ingredient. Exposed to sunlight, the battery powered a small electric motor indefinitely. Professor Colin Fink of Columbia University developed a similar power converter in 1935. This device consisted of layers of copper and copper oxide. Like the others, it remained only a curiosity.

We know the sun itself has been around a long time, but it comes as a surprise to realize that men have been trying for so long to harness its energy. There have been many factors against early success, even when some pioneers must have felt they had it surely in their grasp. First was a lack of knowledge. And when men acquired knowledge they were still short of the technology to apply it. Most frustrating of all, however, has been the apathetic lack of interest by their fellow men in this bounty from the sun.

Those who do not know history are doomed to repeat it.

Anonymous

4 WE DISCOVER THE SUN—AGAIN

Reinventing the wheel is an occupational hazard of science, and nothing typifies this more than the periodic rediscoveries of solar energy. Progress has not continued on a uniform upslope since the time of Archimedes or John Ericsson or Charles Abbot. Instead it has lurched along mainly by fits and starts, mostly to dead-end in frustration and failure.

Historically the use of the sun is linked with the military, beginning with the "signal mirror" effect of sunshine from the polished helmets or shields of soldiers. In mythology the sun has often been called down to destroy; the solar mirrors of Archimedes represent the first human use of the sun as a weapon of destruction.

However, despite Hermann Oberth's suggestion of huge mirrors in orbit about the earth, not only melting ice floes and warming fields for agriculture, but razing whole cities as well, the only practical military application of solar energy until World War II, when the most recent rediscovery of the sun's potential began, was the heliograph. This solar signal was used in British and American military operations of long ago, including the conquest of America's Southwest from the Indians. Signal mirrors were used by downed flyers in World War II, and the sun was put to use in a different way to save aviators' lives.

The Navy's Solar Survival Still

Coleridge's "water, water, everywhere, nor any drop to drink" was brought cruelly home: You can't drink salt water and survive.

FIG. 23. Survival still designed by Dr. Maria Telkes for the U.S. Navy in World War II.

Solar energy came to the rescue in those areas without conventional means of providing fresh water. Dr. Maria Telkes invented plastic stills small enough to pack away in a survival kit, yet able to desalinate about a pint of sea water a day. In operation, the inflated plastic bag is placed on the water and a small amount of salt water poured in. The sun raises the temperature inside the bag sufficiently to cause evaporation and fresh water condenses on the plastic surface. Periodically the fresh water is drained from the still and more salt water added.

Thousands of these stills were produced during the war, and they were standard equipment in life-raft kits. More recently improved models have been developed, including one humorously called a "sit still." When the sun is not shining the still can be warmed by being placed beneath the user like a cushion.

The portable stills of the war sparked some interest on the part of the U.S. Department of Agriculture, and also the Interior Department's Office of Saline Water, in developing larger stills. A number were built in Florida and elsewhere under the impetus of growing shortages of fresh water in many areas. In addition to the construction of glass-covered stills such as were built in Chile, research was begun toward using the cheap, lightweight transparent plastics becoming available.

Solar Homes

In 1952, with the memory of World War II shortages still vivid in their minds, a President's Materials Commission wrote a glowing report on the possibilities of using solar energy to heat houses in the United States. By 1975, the official report stated, there would be a market for about 13 million solar-heated homes. Since fuel for space heating represents about 30 percent of our total energy outlay, such a switch to sunshine would save great quantities of fossil fuels.

As a result of this optimistic dream, a number of institutions and some individuals began experimental solar-heated homes. Some interesting projects were demonstrated, including the MIT home in Massachusetts, those of George Löf in Colorado, one in Arizona, an office building in Albuquerque, and several very successful houses in Washington, D.C., designed and built by inventor-engineer Harry Thomason.

Solar water heaters were as old as the hills, and there were many such installations in southern California in the 1920s and 1930s,

FIG. 24. Winning design in a solar home competition conducted by the Association for Applied Solar Energy. This was built in Phoenix, Arizona.

some of them still functioning today. After World War II a real boom in solar water heaters began in Japan and also in Israel. In fact, heaters by the hundreds of thousands were installed in Japan. Interest spread to Florida, and for a time there were thousands of sun-heated coils providing hot water in many houses and apartments. Solar enthusiasts even succeeded in installing solar swimming pool heaters, and attempts were made to commercialize this venture, since gas heat for pools was very expensive.

The Solar Furnace

Solar energy could heat a lot more than water, of course, and there was a new wave of interest in the idea of the high-temperature solar furnace for use in research and industry. Largely because of the success of a remarkable French scientist named Felix

Trombe, who built a huge, thirty-five-foot diameter solar furnace in the Pyrenees shortly after the war and used it not only for research but for commercially smelting materials in the "pure heat of the sun," small solar furnaces by the dozens began to appear in the United States and in some other countries.

Mostly put together from surplus five-foot searchlight reflectors, these solar furnaces generated temperatures of about 6,000 degrees C. In addition to very high temperatures, solar heat offered a complete absence of contaminating gases, magnetic fields, or other undesirable conditions found in conventional high-temperature arc furnaces. The Air Force seriously proposed construction of a mammoth hundred-foot furnace in New Mexico for high-temperature research, including nuclear blast studies.

The U.S. Army Quartermaster Corps did build a solar furnace almost as large as that of Trombe, at Natick, Massachusetts. On another front, hobbyists learned they could fire ceramic jewelry handily in simple, back-yard solar furnaces.

FIG. 25. Telephone worker connects solar battery power source to system at Americus, Georgia.

The solar cooker, either a reflector type somewhat like the furnace, or a simple glass-plate oven, enjoyed popularity thanks to many articles in how-to-do-it magazines. Inexpensive to buy, or easy to build, solar cookers were prize winners at science fairs in schools across the country. Ovens produced temperatures higher than 400 degrees F., and reflector cookers did a competent job on hot dogs, hamburgers, and even steaks, and could boil a quart of water in about twenty minutes.

Ma Bell's Wonderful Solar Battery

In January of 1954 a brief paper was published in the *Journal of Applied Physics* by D. M. Chapin, C. S. Fuller, and G. L. Pearson, chemist and physicists respectively with Bell Telephone Laboratories. The first paragraph of this document read:

> The direct conversion of solar radiation into electrical power by means of a photocell appears more promising as a result of recent work on silicon p-n junctions. Because the radiant energy is used without first being converted to heat, the theoretical efficiency is high.

This modest prediction proved to be one of the scholarly understatements of our time. The "recent work on silicon p-n junctions" resulted in the solar battery, the most significant breakthrough in the field of photovoltaic cells since the discovery of that phenomenon. Typical efficiency of commercial photocells in 1954 was only about 0.5 percent. The paper written by Chapin, Fuller, and Pearson detailed a solar battery that converted sunlight into electricity at an efficiency of about 6 percent, twelve times as high as earlier cells.

The Bell workers estimated an efficiency of 22 percent as the theoretical maximum for their new device, an outgrowth of work with semiconductors, particularly in the transistor field. Before long they had nearly doubled the efficiency of the early cells and were producing about 11 percent efficiency; they were also proving feasibility in an application providing power for a rural telephone line in Georgia. Interestingly, the efficiency of a small gasoline engine is in the neighborhood of the 11 percent figure.

Bell licensed other manufacturers, who began to manufacture solar batteries in great quantities. Among these were Hoffman Electronics Corporation, International Rectifier Corporation, Heliotek, and several others. Soon solar radios and a variety of hobby kits were being marketed. However, as with the portable solar stills, it was the military that provided most of the market for this new solar device.

Military Interest

The Navy was interested, not only in solar stills, but also in the use of solar battery power supplies for buoys and other installations requiring a self-contained power supply. One advantage of using solar power over possible nuclear packs is the safety factor—an important consideration, since one buoy washed ashore in Nica-

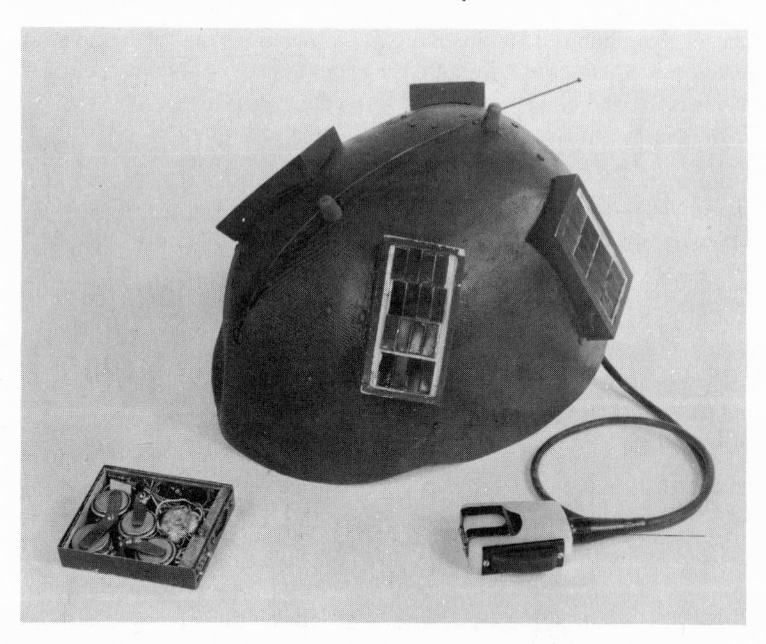

FIG. 26. Solar-powered "walkie-talkie" mounted on GI helmet.

ragua was hacked open with a machete by natives! Fortunately it was powered by chemical batteries.

The Air Force, too, became active in solar energy projects. Small, solar-powered radio transceivers were developed for aviators' survival kits. The advantage is obvious: solar batteries can be stored for an indefinite time with no loss in power, and when called on will continue to function indefinitely. In addition, the Air Force financed continuing studies of solar batteries and other phases of solar energy application. An example was work "farmed out" by the Cambridge Research Center of the Air Force to scientists in Israel.

The Army was interested in solar energy as well, and put it to use in many applications. It was Signal Corps people who produced the pioneer solar radio for Vanguard. Work continues in solar battery research into radiation damage, large-area cells, and so on. On the ground the Signal Corps used large solar panels to transmit radio messages over long distances, and smaller units in helmet radios for soldiers. Lest our troops be fair-weather soldiers, chargeable storage batteries were added.

Solar Science Comes of Age

In 1954 solar science took another step ahead, this one of a formal nature. In that year the Association for Applied Solar Energy was formed in Phoenix, Arizona, with Henry B. Sargent as its first president. AFASE was formed by a group of academic, industrial, financial, and agricultural leaders in the Southwest to stimulate research leading toward greater utilization of solar energy, and the "Valley of the Sun," scene of solar experiments half a century before, was a fitting choice.

Stanford Research Institute and the University of Arizona helped the infant organization get started, and in 1955 AFASE organized the first World Symposium on Applied Solar Energy. Lewis W. Douglas was chairman, and Merritt Kastens of Stanford vice-chairman, with an advisory committee including scientists

Charles Abbot of the Smithsonian Institution, Vannevar Bush, Godfrey L. Cabot, Dr. Maria Telkes of New York University, Frank Lloyd Wright, and other noted experts.

Registering for a conference in Tucson and symposium in Phoenix were representatives of the sciences, education, government, industry, and finance. In addition to scientists from the United States, delegates from thirty-six other countries came to present papers and to demonstrate their projects at the solar engineering exhibit. This high spot of the symposium showed more than eighty solar devices ranging from cookers and steam engines to radios powered by the solar batteries invented only a year earlier.

The AFASE symposium set the stage for a new era of solar energy. The organization sponsored other conferences and symposia, including one in 1958 for youthful scientists; built a solar-heated home; and introduced two publications that detail progress in the solar field: *The Sun at Work* and *Journal of Solar Energy*.

Many research organizations throughout the country were also involved with solar energy projects. Among these were Stanford Research Institute, Battelle Memorial Institute, Arthur D. Little, Inc., National Research Corporation, Franklin Institute, Bjorksten Research Laboratories, Eppley Foundation for Research, General Motors Research Laboratories, Charles F. Kettering Foundation, Mellon Institute of Industrial Research, Minneapolis-Honeywell Research Center, and Yellott Solar Energy Laboratory.

Colleges and universities took on solar research programs, and on some campuses special courses were given pertaining directly to solar energy. The following listing, while not complete, is representative:

Arizona State University and the University of Arizona both engaged in research work, and at the latter institution a solar laboratory was heated and cooled with solar energy.

The University of California's Department of Agricultural Engineering did work on solar-heated buildings and pools. Berkeley's radiation laboratory investigated organic-dye solar batteries, which would be cheaper than silicon, and the College of Engineering worked with solar stills and power systems utilizing the difference of temperature in the ocean, which, of course, is a form of solar energy.

The University of Florida, under the direction of Dr. Erich Farber, embarked on an active solar energy program and did much work on behalf of heating and ventilating engineers.

The Massachusetts Institute of Technology developed solar house heating, and also researched many phases of photochemistry.

Michigan State University's Agricultural Engineering Department conducted studies of solar energy availability, collection, and storage for farm use, as well as studies of roof orientation for best solar energy exposure, and the design of solar air heaters.

FIG. 27. Solar exhibition held in Algeria. The large structure at left center is a solar-heated home.

Minnesota University's Department of Mechanical Engineering worked on the design of solar collectors and the utilization of solar energy for drying of crops.

The New Mexico Institute of Mining and Technology investigated the use of solar furnaces and solar pumps.

New York University's College of Engineering Research studied solar heat collectors, low-cost distillation methods, and the use of solar furnaces in high-temperature coatings.

Purdue University's Agricultural Engineering Department worked on solar-powered fence chargers and solar refrigeration of livestock shelters.

The University of Wisconsin had long been interested in solar energy and its applications. Special projects included work on the design of solar cookers and their testing in the field, absorption cooling, solar stills, and energy storage methods. Dr. Farrington Daniels of Wisconsin was one of the world's authorities on solar energy, and author of *Direct Use of the Sun's Energy*.

Around the World with the Sun

One very reasonable idea suggested early in the game was that, if solar energy was important to anyone, it was important to those living in underdeveloped areas of the world, where about the only form of energy available is solar energy. Solar batteries came along just at the time when the United States was embarking on ambitious programs of helping such countries.

Solar-powered radios became a reality in remote areas, not only in the United States, but in the Amazon jungles and in Africa as well. The International Cooperation Administration sent solar-powered "listening centers" to small communities in Paraguay, and planned the same program for India, Pakistan, and Afghanistan.

The United States Agency for International Development ordered one thousand transistorized television sets suitable for solar power for installation in remote areas (incidentally getting

FIG. 28. Solar power plant designed by Russian workers. Large mirrors mounted on railroad cars on circular track were to have focused sunlight on the central boiler.

itself accused of global boondoggling). Another application of solar energy was to power a seventeen-foot collapsible boat designed by AID for transportation on the jungle rivers of Surinam in South America. In this intriguing application solar batteries formed a sun shelter on the boat and provided two hundred watts of power for a small electric trolling motor to move the boat along at three knots. The solar batteries also provided power for radio communication and for operating a small electric drill.

Of course, the United States had no monopoly on solar energy research, and work was going on in several nations that was as far advanced as ours, and in some cases even more sophisticated. With the postwar stimulus in a variety of fields, solar science flourished in Europe and Asia. In addition to the leaders in the field, many other countries had at least token programs, and the list of solar activities internationally read like the alphabet.

RUSSIA. Dr. V. A. Baum guided Soviet research as director of the Solar Energy Laboratory of the Power Institute of the U.S.S.R. Academy of Sciences. Originated in 1950, the Heliotechnical Laboratory at Tashkent turned out precision reflectors used for welding, and solar cookers the equivalent of a six-hundred-watt hot plate. One clever device was a solar boiler that folded up compactly into a suitcase. Water heaters of all sizes were developed and used not only in Tashkent but in Ashkhabad and Tiflis as well. The most ambitious was a large heater in Tashkent, which boiled 500 liters of water daily. In the capital of Uzbek, a solar boiler produced 130 kilograms—about 300 pounds—of ice a day.

The Russians designed a large solar boiler to be built in Soviet Armenia. This plant was to produce 2.5 million kilowatt-hours of

FIG. 29. Huge solar furnace at Odeillo in southern France. The curved mirror is built into the multistory laboratory building.

electricity a year, but since it required some five acres of mirrors to reflect sunlight onto the boiler, only a pilot plant was built.

Soviet scientist Abram Ioffe sparked the revival of thermoelectricity in the 1930s, and Russia built solar-powered thermoelectric converters. For example, using a six-foot mirror to focus heat onto a semiconductor thermoelectric generator, they claimed to have produced one-horsepower engines.

FRANCE. Under the leadership of Dr. Felix Trombe, France continued its leadership in the furnace field. The big thirty-five-foot solar furnace at Mont Louis remained the largest in the world, and surely the most active and productive, since it operated continually both for research and industrial smelting. Not resting on his laurels, Trombe began a truly huge furnace at Odeillo, including in his plans a mirror 115 feet by 165 feet, which would actually form the side of a laboratory building. The Odeillo furnace was to have an output of one thousand kilowatts, making it the first solar installation in the megawatt, or million-watt, range.

ISRAEL. Inspired perhaps by their desire to make the desert "bloom as the rose," Israeli scientists were doing some of the most interesting work in the solar field.

Headed by Dr. Harry Tabor, workers at the National Physical Laboratory of Israel produced five-horsepower turbines "fueled" by inflated plastic reflectors, highly efficient heat absorbers for use in heaters and power plants using solar energy, and a promising idea called the "solar pond," in which a large pond was treated chemically to store the heat of the sun, then covered with plastic. After a certain generating period, heat stored in the water was used to power a steam engine. Since this type of energy collection and storage is cheap and simple, it was felt that there were great possibilities for the solar pond idea.

The Israel Institute of Technology developed water heaters, coolers, stills, and dryers. The leading solar energy product of Israel

FIG. 30. Pilot model solar-powered turbine built by Israeli workers led by Harry Tabor.

was the "Miromit" solar water heater, a unit so efficient that the country's electric utilities were forced to drop their prices to compete with the sun, perhaps the first time that solar energy had been economically competitive with conventional power on Earth.

Heaters were produced at the rate of four thousand a year, and sold in Israel and four other countries.

The Israeli scientists were also doing work for the Cambridge Research Center of the United States Air Force on "selective surface" heat absorbers. Dr. Tabor and his colleagues developed the most efficient blackened collectors known for trapping solar heat, capable of storing from 80 to 90 percent of the heat striking them.

JAPAN. The Japanese were also active proponents of solar energy; the Kobayashi Institute of Physical Research in Tokyo and the Government Industrial Research Institute in Nagoya were two centers of activity. Japan was progressing in electronics, and her scientists invented such advanced semiconductor devices as the revolutionary tunnel diode. It was not surprising to find production of solar batteries among their accomplishments. These were put to practical use in remote radio installations and in lighthouses, where they function for long periods with little maintenance.

The Japanese people love baths, and in Japan the solar water heater was mass produced. More were used there than in the rest of the world combined. By 1960 there were 250,000 in use, with, an estimated annual saving in coal of one ton per heater, and with heaters ranging in price from nine to eighty dollars.

Solar cookers were also produced in a variety of shapes, and there were industrial and research solar furnaces in operation. A more unusual solar energy application was the growth of algae in plastic tanks. The Japanese, critically short of food, had done more work in this direction than any other country, even producing chlorella food supplement powders commercially. An acre of pond area produced about twelve tons of algae yearly.

Other research and development projects ranged from the use of solar heat to distill sugar-cane juice to the heating and cooling of residences. One clever invention combined the inflatable life preserver with a solar still!

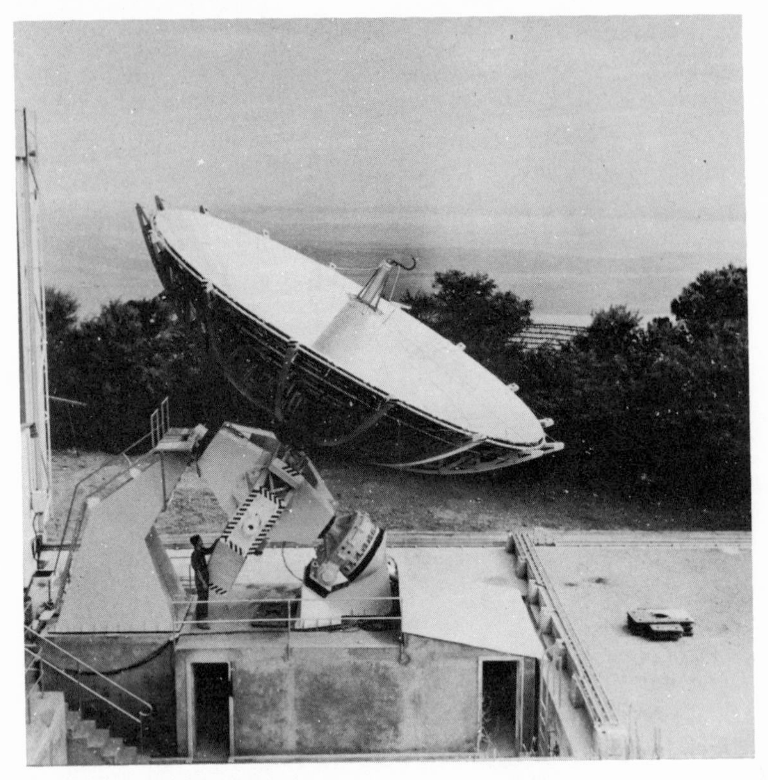

FIG. 31. This large solar furnace at Bouzareah, Algeria, was used to fix nitrogen from the air.

OTHER COUNTRIES AND THEIR CONTRIBUTIONS. *Algeria* was active in solar energy research perhaps as a defensive gesture, since much of her territory lies in the Sahara Desert. One logical goal was the air-conditioning of desert houses using solar heat. The third largest solar furnace in the world was located at Bouzareah, used for photochemical research and for producing nitrogen from the atmosphere. Also under study were thermoelectricity, selective surfaces, and solar batteries. "Radiosol" water heaters were produced commercially.

Australia had formed an active branch of AFASE, and one

development was the installation of a twelve-foot solar furnace on the Kensington Campus of the University of New South Wales. The furnace developed about five kilowatts of heat, and it reached temperatures hotter than 3,300 degrees K. Production of salt from sea water was another solar energy application, and experiments were made on spraying brine through the air to speed evaporation. The World Power Conference held in Melbourne in 1962 featured many papers on solar energy.

Burma built low-cost cookers, heaters, and stills, and experimented with the use of solar energy in salting fish. Scientists suggested that a solar-powered refrigerator would be of more value to the Burmese than a cooker.

Canada built a solar-heated house, despite its far-north location. In search of a warmer climate for its solar researchers, McGill University opened the Brace Experiment Station in Barbados, West Indies.

Ceylon, sweltering in the heat, also researched the use of solar energy for refrigeration. In 1955 scientists there performed a valuable service by measuring solar radiation during an eclipse.

Chile was the scene of large solar stills in the last century, and engineers again proposed such installations. The sun was being used in evaporating pits for chemical companies, and research was being done on heaters, cookers, furnaces, and solar batteries.

China, both Nationalist and Communist, was active, like Japan, in the solar energy field. On Taiwan the population had doubled since 1946, and estimates were that local fuel resources would last only another forty years at present rates of consumption. Therefore, work was being done at Taiwan Normal University to produce cookers, space heaters, water heaters, and furnaces for island use to save conventional fuels.

On the mainland, it was reported that about eighty Red Chinese factories, mostly in Shanghai, were turning out heaters and cookers.

Egypt. It had been here, back in 1913, that Shuman and Boys had set up the world's largest solar steam plant, rated at one hundred horsepower. Now scientists were investigating sea water distillation, solar heating, drying, baking, pumping, and so on.

England, though quite far north, had done a fair amount of solar research. Many of its scientists were pessimistic about the possibilities, but Professor Harold Heywood was quite active, and worked with Theodor Finkelstein on the early models of a revolutionary hot-air engine that could run on solar energy.

Germany had a few exponents of solar energy. Dr. J. H. Dannies had been actively experimenting with solar heating, cooling, drying, and even extracting drinking water from the air with solar energy. As early as 1935 he built a solar refrigerator, and tested improved designs in the Negeb Desert in Israel. Some algae culture research had also been done.

India had excellent solar scientists in Dr. M. L. Khanna and others, and had developed practical solar cookers, stills, and heaters. With the National Physical Laboratory at New Delhi as headquarters, work was going forward on solar refrigerators and solar power plants for villages.

Italy, host to the United Nations energy conference in 1961, built water heaters and space heaters, among other projects, and residential installations of water heaters were made. The University of Bani was a research center.

Lebanon's Adnan Tarcici perfected a collapsible solar cooker. A onetime United Nations delegate, Tarcici patented his cooker in the United States.

South Africa continued research on space heating, water heating, stills, and solar furnaces.

Spain had a "Special Energy Commission" that carried out work in solar energy research.

Switzerland. Swiss watchmakers in the Advanced Research Division of Patek Philippe Company were manufacturing and marketing solar-powered clocks to bring the ancient sundial concept of time from the sun up to date.

The United Nations had long been interested in the development of solar energy, particularly with respect to applications in the underdeveloped lands of the world. In 1961 the U.N. spon-

FIG. 32. Mass-produced solar cooker experimented with in India. Such simple stoves can produce about five hundred watts at the cooking surface.

sored the United Nations Conference on New Sources of Energy in Rome. Representatives of some sixty countries were on hand to hear papers by specialists in thirty countries on the use of solar energy, wind power, and geothermal power. Solar energy was a main concern of the conference, and of particular interest were the revolutionary Battelle Institute Stirling-cycle solar engine and the five-horsepower solar turbine from Israel.

The Failure of Nerve

By the mid-1950s it seemed that the solar energy enthusiast's cup was about to run over. In 1955, when the infant Association for Applied Solar Energy held a worldwide solar symposium attended by one hundred and thirty scientists from thirty-seven countries, most people were amazed at what could be done with solar energy. Solar scientists and industrialists left the meeting—which had demonstrated engines, stills, furnaces, and a variety of cookers and heaters—highly encouraged. Stories flooded the press and the popular magazines, touting the wonders and great potential in solar energy. It seemed too good to be true—and it was. There is an old saying that goes "when thy cup runneth over, looketh out." Sure enough, the solar bubble burst once more.

Unfortunately, much of the publicity was of a "gee whiz!" nature, and did more harm than good. Fanciful drawings of homes roofed with solar batteries and autos rolling along on free sunshine were taken too literally, and when people learned that such a roof would cost hundreds of thousands of dollars and that the solar car was as far away at the moment as antigravity screens, interest understandably waned.

A few brave men still had the conviction that our future yet lay with solar energy. A handful of solar scientists and engineers continued to work, and a few interested people in government poked and prodded to get government funds for various projects. In 1959 Senator Alan Bible of Nevada, a state that reckons its sunlight an asset second only to its slot machines, proposed setting up a research fund of $10 million for solar energy. Introducing Senate bill 2318, Bible said in part:

> Some one once calculated that the enormous power involved in the huge atomic energy reactors at Hanford, Washington, does not surpass

the solar energy, which every day, in the form of sunlight, falls unused upon the roofs of the Hanford buildings. To express it another way, there is more energy received from the sun on a half square mile on a sunny day than there is in one pound of U 235.

The bill was duly referred to the Committee on Aeronautical and Space Sciences, the group that seemed to come closest to sun power. No favorable action was taken. Senator Hubert Humphrey of Minnesota later introduced a similar bill, which had about the same luck.

The Association for Applied Solar Energy weathered the solar storm. Housed at Arizona State University, and with funds from the National Science Foundation, AFASE continued to act as a clearinghouse for information and to stir up interest for further conferences on solar energy.

Meetings continued to be held and papers to be written, adding to the body of knowledge being built up on solar science. Ironically, much of the effort occupying engineers was not the harnessing of the sun's power but the rejection of its heat. Heat is, of course, energy, and each new method of foiling the entry of this heat is actually a wasting of energy. To cool our homes in summer we burn up precious fuel to dissipate solar energy, which would be better put to use to power the cooling mechanism.

Another broad area of solar research was simply that of pinning down more exact values for the solar constant, and other quantitative analysis. The study of solar flares, those vast outpourings of radiation which not only confuse communications on Earth but endanger astronauts in space as well, proceeded carefully at Air Force laboratories at Sacramento Peak in New Mexico and other sites.

All this diligent research was fine, of course, but it was seldom reflected in practical applications. Progress was so slow as to be almost imperceptible outside the research laboratory. The journals published scientific articles, but interest seemed to be at an academic level, with little or no hope of ever being reduced to practical application.

Alongside the glamorous and upcoming atom, solar energy was a drab country cousin. To dim its chances further, along came the hydrogen bomb and the suggestion that man could duplicate the process of the sun in such a device. Fusion was hailed as far more wonderful than fission, though some thoughtful scientists suggested it might be better to keep the fusion power plant and its multimillion degree temperatures at a safe distance, say 93 million miles. The fuel for this exotic power plant is deuterium, or heavy water, and everyone knows that the earth is mostly covered by water anyhow. Our problems were solved; surely fusion would provide for us as far into the future as anyone could see.

It was the Nuclear Age that scuttled the hopes of solar energy proponents. It would take the coming of another age to revive the sun's prospects. This was the Space Age, a turning away from the microscopic innards of the atom to the astronomic vastness of the solar system. And man had to take note of the sun again.

Eyes, though not ours, shall see
Sky-high a signal flame,
The sun returned to power above
A world, but not the same.
Cecil Day Lewis, "The Magic Mountain"

5 SOLAR POWER FOR THE SPACE AGE

Literally from a clear sky came the boost that solar energy sorely needed. The exploration of space lifted off from the pages of science-fiction magazines and became reality: Russia lofted Sputnik I to open a new and challenging era. The United States followed suit after an embarrassing lag, and space was soon sprinkled with dozens of satellites and probes. It was shortly evident that once in orbit, or on their way through space, such vehicles had need of a power supply and the solar battery, still too expensive for earthbound applications, rose to the occasion like a champion. In a modest debut, several solar batteries of an early type were mounted on the Navy's Vanguard I, a crash attempt at a space vehicle.

Solar Batteries: The Voice from Space

Working feverishly with the Navy scientists, United States Army Signal Engineering Laboratories (USASEL) installed a tiny radio, powered by six small solar panels, on the Vanguard TV-4 (for Test Vehicle 4). This project was already plagued by several failures of the booster, and there was much criticism of all the services trying to launch their own satellites instead of consolidating efforts and saving time and effort. But on March 17, 1958, Vanguard re-

FIG. 33. The small Vanguard satellite that was our first space success. The solar cell array is in the center.

warded its backers and confounded the critics by placing a satellite the size of a grapefruit and weighing only three and three-quarter pounds into elliptical orbit ranging from about four hundred miles to more than twenty-four hundred miles above the Earth. USASEL received a congratulatory teletype message from the Navy almost immediately:

> EASTON TO ZAHL, ZIEGLER, HUNRATH, HERCHAKOWSKI OF SIGNAL ENGINEERING LARORATORIES. CONGRATULATIONS! SOLAR-POWERED TRANSMITTER WORKING PERFECTLY. YOUR COOPERATION ON THE PROJECT DEEPLY APPRECIATED.

How perfectly, and for how long, the solar-powered transmitter would continue to work could not have been predicted by the most highly elated of the crew. Nor was the full implication of the feat yet apparent: solar batteries by the hundreds of thousands would soon follow the 108 tiny silicon chips mounted on Vanguard I's six-inch satellite.

Vanguard was well named. While its tiny solar power plant had a designed output of only a tenth of a watt, hundreds of later satellites carried ever larger and more sophisticated arrays of solar batteries. In the United States other space vehicles followed, each of them bigger and mounting more elaborate equipment and instrumentation than the one before. As an example of the popularity of solar batteries for space use, by 1965 NASA was using close to a million a year. This did not include those used by other countries. Russia's Sputnik III incorporated solar batteries, as have most of her satellites sent aloft since.

Even before space vehicles had proved feasible, scientists and engineers were lying awake nights thinking up other tasks for them besides idly sweeping the sky in their orbits. The "eye-in-the-sky" has long been a provocative idea, and such an Orwellian exploitation of space was a natural development. While early satellites were scrupulously shot into orbits that would not lend themselves to such cloak-and-dagger motives, it was inevitable that projects like one dubbed "Big Brother" be developed. Midas and Samos and Vela Hotel are some of the colorful names of spy craft that have been lofted. All such orbital sentries have one requirement in common: they need a constant source of power to monitor atomic explosions, missile launchings, nuclear warheads, and so forth. Solar batteries offer that source, plus freedom from any radiation of their own that might confuse the delicate monitoring equipment.

Another eye in the sky of a more creative nature is the weather satellite, and the proposal was made for a spacecraft mounting a photocell to scan the Earth. This simple device would transmit either a spot of light or nothing, depending on the presence or

FIG. 34. A Tiros III weather satellite covered with solar cells to provide power for camera drive and other needs.

absence of cloud cover. By putting together enough of these white and dark spots, it was hoped that weathermen might get a rough idea of what Earth's weather pattern looked like from several hundred miles up. The scheme was thought overly optimistic by many, but even before it could be put into practice space technology leapfrogged it, coming up with a weather surveillance satellite that made the light- and dark-spot method as crude as the earliest moving pictures compared with today's three-D, wide-screen, Technicolor epics. This was Tiros, the weather eye, made practical by the solar battery.

For many years astronomers, physicists, and other scientists have dreamed of observatories free of the dust and haze of Earth, and unhampered by the disturbance of its atmosphere. Until the orbiting satellite the best that man could do was to build his research facilities high atop mountains. This helped, but it was far from an ideal situation. What was needed was an observatory out in the vacuum of clear, uncontaminated space. There its instruments would record without the degradation caused by atmosphere, artificial lights, and terrestrial magnetism. This dream is now a reality, and solar-powered observatories study Earth and the celestial bodies of interest.

FIG. 35. Drawing of Skylab, to be orbited in 1973. Solar panels produce the equivalent of about fifteen horsepower for duration of the mission.

An early space observatory called the Orbiting Astronomical Observatory was an unmanned satellite measuring seven by nine feet and fitted with two solar panels providing about 250 watts of electric power. Such pioneering efforts have now culminated in Skylab, a twenty-eight-ton structure in which three astronaut scientists will spend weeks in space. Equipped with a large telescope, Skylab mounts extendable solar panel arrays that produce kilowatts of electricity to power the variety of instruments aboard. Beyond Skylab is a full-scale space station capable of housing one hundred or more scientists, physicians, researchers, technicians, photographers, and other specialists. Again, solar batteries will provide the electric power to operate the space station.

Big Switchboard in the Sky

One of the biggest and most important jobs that solar energy plays in space is that of powering communication satellites like AT&T's fabulously successful Telstar. Hermann Oberth first suggested a crude form of message relay in 1920. British science writer Arthur C. Clarke advanced the idea of satellite TV in 1945, and Dr. John R. Pierce, of Bell Laboratories, was first to put satellite communication into practice. When NASA announced plans for lofting a big hundred-foot plastic balloon into space, Pierce suggested experimenting with the bouncing of radio waves from it. Echo, as it was called, was a "passive" repeater station. To be practical, a satellite repeater needs a power supply to boost the faint signal it receives to a level that will carry it back down to the receiver strongly enough for economical reception.

As Bell Telephone Laboratories pointed out, there are no telephone linemen in space; an active repeater satellite must be highly reliable. Development of Telstar cost more than $50 million; added to this was the cost of the launch by NASA with a big Delta booster, another $3 million. With an expenditure of this size Telstar had to be designed for years of unattended service. There was only one way to provide this reliable lifetime—the use of solar batteries.

FIG. 36. Telstar communication satellite, which introduced a new means of radio and television transmission around the world.

AT&T was not the only group active with communication satellites, of course. Among the others was RCA, with its Relay, a similar, 172-pound repeater. Hughes Aircraft also entered the field, and its concept was a departure from that of Telstar. Most satellites move around the earth in their orbits; Hughes' 80-pound

Syncom satellite was a "twenty-four-hour," or synchronous, satellite.

Syncom flies an equatorial orbit more than twenty-two thousand miles high, and appears stationary in the sky. There are important advantages to such a system. Because they are fixed and high, three Syncoms cover the earth for communications. And radiation harmful to solar batteries is less at this altitude.

Today Comsat, the communication satellite, is a fact of life, and brings instantaneous coverage of events all over the world. For this service alone the solar battery earns its keep.

Solar power was not long limited to orbital flight. Soon space vehicles were departing the earth on trips to the moon and the planets. Pioneer V was launched to probe deep into space, and its solar-powered radio sent back data from more than 22 million miles out. Even this was but a prelude, and the Mariner Venus probe used the sun to send back information from a distance of 36 million miles.

As its space programs began to unfold, the Air Force became increasingly aware of the problem of communication in space. As part of its research program, Wright Air Development Division gave a contract to Electro-Optical Systems, Inc., to investigate the possibility of using solar communication, in effect a Space Age heliograph. The result was SOCOM (for solar communication), a system that demonstrated the feasibility of using a light beam to transmit messages through space for distances up to 10 million miles.

Oberth's fantastic scheme of "death-ray" mirrors hung in space has been ridiculed by other writers, who have different ideas on how best to use mirrors. It has been suggested that they could be used to light up cities at night, one expert estimating that a thirty-meter mirror could provide as much light for a town some twenty kilometers across as would a full moon (not to consider the romantic aspects of such an artificial moon!). Disregarding the wilder suggestions, it is obvious that there is much magic in mirrors, and that they can do many things with sunlight.

Whether we continue to use radio frequencies for space communication, or change over to those of visible light or infrared, remains to be seen. In all probability both types will be used, depending on the job at hand and its requirements with regard to distance, message-handling capacity, security, and so on. Whichever way communication goes, however, the sun is sure to play a big part in the process, and solar power will be used increasingly to speed the word along.

Power for the Space Age

The spectacular success of solar energy in providing power for communication exceeded the fondest hopes of early space program planners. So great was their success, in fact, that proposals were soon made for using the sun to provide power for other uses aboard space vehicles. There are two excellent reasons hastening the development of practical power supplies in space: the increasing power demands of space vehicle equipment, and the increasing duration of the missions themselves. By the time the first flush of success over early Russian and American satellites wore off, the deficiency of their conventional power supplies was apparent. Satellites orbiting earth silently were not only eloquent tributes to the skill and energy that put them there, but also mute testimony to the need for power supplies as long lived as the satellites themselves.

As the weight of unmanned satellites and space probes increases to hundreds of pounds, and in some cases to tons, power requirements are likewise measured in hundreds and thousands of watts. Manned capsules make even heavier demands on power, including those for heating and cooling, lighting, pressurization, operation of controls, and experiments.

One of the hardest facts of life in space is that it costs many pounds of expensive rocket fuel for each pound of payload placed in orbit or launched as a deep-space probe. In other words, to put a pound of batteries in orbit might require more than one hundred pounds of fuel, fuel that has weight itself and uses up much additional power with no useful return. If we weigh a conventional fifteen-kilowatt generator, with sufficient fuel to run it for a year, and calculate the size of a rocket big enough to lob it into space, the results are discouraging. This was the reason for all the current research into solar power plants to accompany new and sophisticated space vehicles aloft.

There are inherent reasons for the spectacular success of solar power in space. One is obvious: we use solar energy in space for the same reason men climb Everest—it is there. In fact, the higher we climb, the more solar radiation we can make use of. On Earth there are roughly one thousand watts of available energy per square meter of exposed surface. In space this figure increases by 40 percent. Thus the solar panel that develops one thousand watts at sea level might well deliver fourteen hundred watts in orbit.

However, while the solar battery had made its mark as a supplier of power in small and moderate amounts, beyond a few kilowatts output there were serious problems of cost as well as collector design. The time had come to consider a different kind of solar power plant, literally a steam engine for space.

Steam Engines in Orbit

The solar "paddle wheels" on the Pioneer satellite have sometimes given the erroneous impression that the sun's energy was actually churning the probe through deep space, much like an up-to-date version of a Mississippi side-wheeler. Although an even more fanciful concept may some day propel space vehicles, solar paddle wheels unfortunately do no paddling at all. In fact, solar batteries may be too expensive to be used as propulsive power for any sizable spacecraft.

It is true that many thousands of solar batteries have been stacked up to provide hundreds of watts of power for communication, driving motors, and the like. But to generate kilowatts of power in this manner will be very costly. A proposed four-kilowatt power plant of solar batteries would cost millions of dollars, and as

FIG. 37. Pioneer V, with its "paddle-wheel" array of solar cells.

we progress to greater power the necessary panels quickly become not only priceless but unwieldy as well. As space vehicles became bigger and more sophisticated, with greater power requirements, designers turned to other solar energy devices than batteries.

A simple, economical solar reflector can deliver as much as half the available energy to its focal point. This represents an efficiency of 50 percent; and better reflectors, more than 80 percent. Eighty percent of the available fourteen hundred watts of solar radiation in each square meter represents more than a kilowatt of power. And a kilowatt, remember, is more than one horsepower. Of course, unless we are heating a house, or frying eggs, heat energy is not particularly attractive to us; we must convert it into some other more practical form. This is generally electricity. And today the turboelectric generator, using heat as a source, is the most efficient converter in use, running at rates that in some cases approach 40 percent. Thus, with heat collector efficiency of 80 percent and electrical conversion 40 percent of that, we still end up with about one-third horsepower for each square yard of collector area.

The first practical proposal for a solar space power plant was put forth by an Austrian named Noordung. This gentleman proposed a space station some twenty-two thousand miles from the Earth in a twenty-four-hour, or stationary, orbit. His ideas were actually predictions, and included wheel-shaped, slowly spinning living quarters to provide artificial gravity. He also designed a power plant in the form of a parabolic mirror focusing the sun's rays onto boiler tubes. Electricity was to be generated for immediate use, and for storage during dark periods. Since this concept was advanced in 1929 we must credit Noordung with much foresight. More recently, space authority Wernher von Braun designed a space station for the Mars Project concept. His book, published in 1952, describes a wheel-shaped vehicle, also with a large parabolic mirror for concentrating sunlight on a boiler.

With the exception of the vacuum and no-gravity conditions of space, and the means of heating the boiler, the design of space power plants, like that of terrestrial applications, is a straightforward job for the thermodynamicist and mechanical engineer. Some years ago government agencies contracted for research and development work on a number of thermal-mechanical plants to furnish power for space vehicles, as it had been demonstrated that these were going to work and would need a source of electrical power.

Prior to 1960 the Tapco Group of Thompson Ramo Wooldridge, Inc., had proved the feasibility of a unit called SPUD, a one-kilowatt solar power unit demonstrator. The prototype ran successfully for 2,500 hours, and TRW got another contract for Sunflower, a three-kilowatt plant that could fit into Saturn and other boosters. The concept featured an unfurling thirty-two-foot reflector, a mercury boiler, and a Rankine-cycle turbogenerator.

Hot Air and Sunshine

The idea of running an engine on sunbeams is apt to be dismissed as a lot of hot air, and in one case, at least, this is a very accurate appraisal. One of the most advanced of the space engines that has yet been built is based on the invention of Robert Stirling back in 1816. General Motors' Allison Division, being knowledgeable in this field, was asked to produce a space engine.

The Allison space engine was designed to deliver the same amount of power as TRW's Sunflower. For use in space the Stirling engine underwent some changes from its land-based counterparts. First, the engine had to be sealed for operation in the vacuum of space. In addition, the working medium would not be air, but helium or perhaps hydrogen. In place of the Sunflower's thirty-two-foot parabolic mirror, the Allison engine relied on a nineteen-foot flat Fresnel lens reflector, which is more easily folded for launching and not subject to the distortion errors likely with a large curved mirror. The smaller reflector was possible because of the increased efficiency of the Stirling engine over the more conventional boiler and generator in Sunflower.

The Stirling engine weighed 567 pounds; it was somewhat lighter than Sunflower. Since it had moving parts, it was subject to the same lubrication problems as other dynamic engines, and also to the effects of no gravity on the expansion and circulation of its working medium.

These sizable solar engines were modest compared with another developed for the Air Force by Sundstrand Aviation's Denver Division. This solar thermal-mechanical giant would produce fifteen kilowatts of continuous electrical power at 120 to 208 volts. Besides the ambitious size of this engine there was another noticeable difference between it and the smaller three-kilowatt designs. Where the Sunflower and Stirling space engines weighed 700 and 567 pounds respectively, Sundstrand's engine weighed, not five times that amount as might be expected with the power increasing by that factor, but only 1,000 pounds. This meant a kilowatt of solar power produced at a weight of about 67 pounds.

An idea of how much power fifteen kilowatts represents can be gained from studying your electric bill. A typical home might use about fourteen hundred kilowatt-hours a month, or two con-

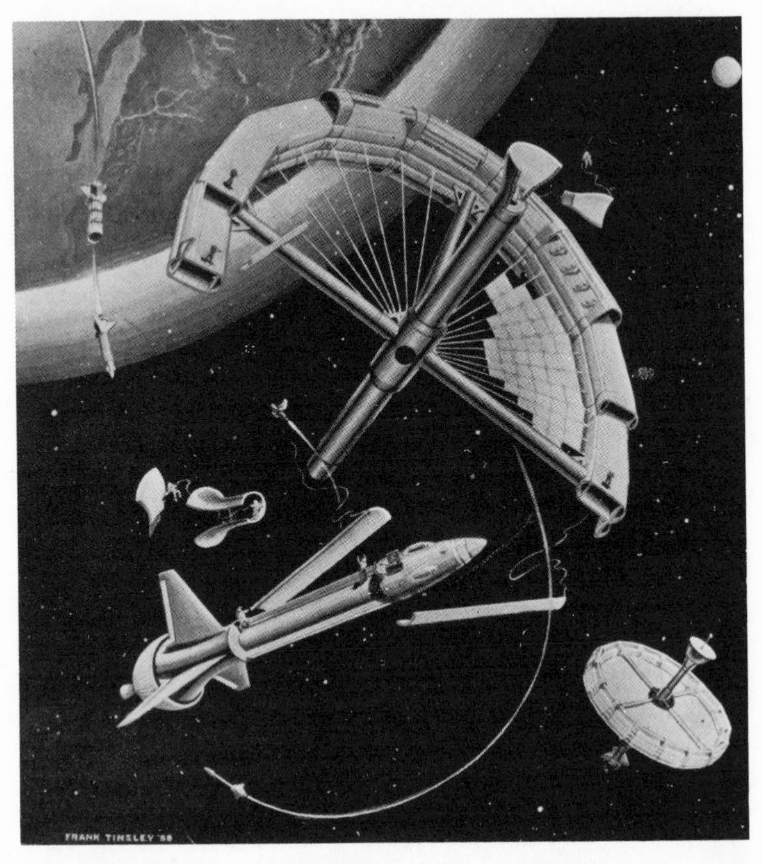

FIG. 38. Artist's drawing of assembly of a large space station with a solar power plant.

tinuous kilowatts, so a fifteen-kilowatt plant would serve seven or eight homes on your block. Obviously a fifteen-kilowatt power plant in the sky can deliver power for a good many uses.

Space Flight on Sunbeams

So far we have considered solar energy primarily as a source of power for communication equipment, light, heat, cooling, and other nonpropulsive tasks for astronauts. There were many who felt that tapping the sun for even these lesser uses was straining the sun as well as the credulity of the public. With the repeated successes of solar batteries in space, generating kilowatts of power for years without replacement or maintenance, the skeptics were forced to reconsider. Nevertheless, the belief of proponents of other sources of energy continued that while the sun may power a radio or cook a meal for an astronaut, it certainly will never propel him through space.

Such a conservative outlook is understandable. Even though Nobel Prize winner Svanté Arrhenius suggested that life was wafted to Earth originally on beams of light, few believe that man today will depart from Earth on that same motive power. Basking in the sunlight, it seems inconceivable that these gentle rays could drive a spaceship weighing many tons on a journey to the moon, much less to the more distant planets.

FIG. 39. Spacecraft equipped with large solar power plant.

The notion of space flight is quite old, the first lunar voyage having been described in Lucian's *True History*, written in A.D. 120. A variety of propulsion methods have been used in fanciful tales that followed, including even wild swans, but it remained for Cyrano de Bergerac to power one of his spaceships with solar energy. His *Voyage dans la Lune* and *Histoire Comique des États et Empires du Soleil* were written about 1650, and while they are as unscientific as might be expected of a man living before the discoveries of Sir Isaac Newton and James Watt, the concept of solar energy is startling. Cyrano also powered one of his craft with rockets, apparently covering all possibilities for space flight.

Three centuries later, almost to the year, other men began to write about space voyages using power from the sun. But these men were not writers of science fiction. In 1954, William Conn,

FIG. 40. Solar-fired spacecraft designed by Krafft Ehricke.

who had built a number of small solar furnaces for laboratories and industrial plants, wrote a short note in a scientific journal suggesting the use of a solar collector to focus heat for detonating the rocket fuel of a missile. Two years later Krafft Ehricke, a German rocket scientist who had come to the United States, delivered a paper titled "The Solar-Powered Space Ship" at a meeting of the American Rocket Society.

Cyrano had journeyed to both sun and moon. Ehricke was more scientific in his approach and more modest in his goals. The solar-powered craft, he felt, was best fitted for circumlunar or cislunar flights, that is, for trips around the moon and in the space between it and Earth. Furthermore, Ehricke's solar-powered spaceship looked more like a balloon, or rather, two 128-foot balloons.

Ehricke's rocket engine would carry 11,000 pounds of liquid hydrogen fuel. A conventional chemical engine would require an additional 50,000 pounds of oxygen to be mixed with the hydrogen to drive the spaceship. But in the solar engine, the sun would heat the hydrogen and produce the driving force. Instead of 25 tons of oxygen, there would be only the weight of the solar "collectors," a mere half-ton in Ehricke's clever solution to the weight problem.

While Ehricke suggested his ship primarily for short trips such as to and from the moon, another scientist from Germany had bolder ideas. NASA's Dr. Ernst Stuhlinger proposed a different type of solar-powered craft capable of voyages through outer space. This guided missile expert designed a much larger ship propelled by electrostatic force! Tom Swift couldn't have imagined such a craft.

The electrostatic drive uses materials like rubidium and cesium as fuel. When these strike an incandescent platinum surface, they cause an ion flow that is accelerated by a negatively charged electrode to create an electronic jet or blast. This emission creates a tiny force similar to the flow of ions in the television picture tube, and is called electric propulsion by scientists and nonsense by the uninitiated.

Although it would have even less acceleration than the Ehricke design, an electrostatic spaceship offered tremendous weight-saving over conventional rocket power, particularly on a flight longer than to the moon. To demonstrate this, Stuhlinger outlined a flight to Mars. His solar-powered craft would carry ten men and a load of 50 tons with a gross weight of only 250 tons. A conventional rocket ship would weigh 1,100 tons. Even more attractive is a much longer flight requiring two years to complete. Stuhlinger's ship would then weigh only 275 tons, as against 7,500 tons for a chemical rocket.

Electrostatic or ion engines have been produced in the few years since Stuhlinger first suggested the revolutionary propulsion means. In 1961 Republic Aviation delivered a variation on this

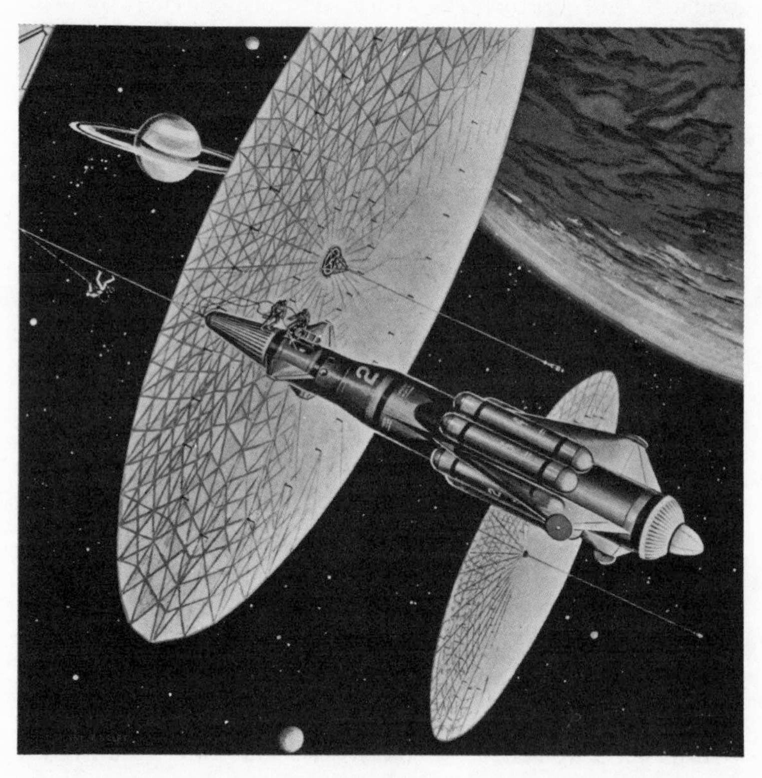

FIG. 41. Space travel using large "solar sails."

theme to the Air Force for evaluation as a space power plant. Not an arc jet or an ion engine in the strict sense, the Republic design was a "plasma-pinch" engine, in which a magnetic field is induced in a hot gas. Powered by a one-kilowatt solar battery supply, the plasma engine developed only one-hundredth of a pound thrust. But with a high efficiency and the long life of its electric power supply, even this feeble engine was a step toward Stuhlinger's dream.

Solar Sailing

Even hydrogen drives, electrostatic engines, and other sophisticated space propulsion systems pale when compared with what is perhaps the most provocative use of sunshine ever proposed—that of "solar sailing."

It has been known for a long time that sunlight exerts a physical force. That force is something like one pound per hundred acres, however, and it was not until 1951 that it was mentioned seriously as power for space flight.

Solar sailing was first suggested in *Astounding Science Fiction*, but it moved quickly into the learned journals. Scientists at MIT, Sandia Corporation, Lockheed Missiles and Space Division, RCA, Westinghouse, Space Technology Laboratories, the University of California, and elsewhere in the United States and abroad, proved mathematically that solar sailing was not only possible but highly desirable. Just as the electrostatic drive is better than chemical rockets, so the solar sail is better than the electrostatic, because it has a higher work capacity per unit weight.

When the idea was first broached, hearers were prone to assume that the speaker himself had stood too long in the direct rays of the sun and that photons striking his head had affected his brain. Vindication of the possibility of solar sailing came with the displacement of Vanguard I by the effect of sunlight. The tiny six-inch sphere, hardly an efficient sail, was moved about a mile in a year, according to scientists keeping track of its orbit. A mile a year seems a pitifully small distance, but sail proponents point out that Vanguard's performance at least proves the theory.

Solar sailing is roughly comparable to sailing a boat on water or ice, or a windwagon across the prairie. On Earth, molecules of air strike the sail and Newton's laws and aerodynamics take over. In space there is no air, but Newton's laws have stood up quite well when put to the test. If a photon of light strikes a solar sail, some of the energy in the light is transferred to the sail. While this seems something like trying to drive a golf ball with a flashlight beam, the basic theory is sound and the implications are of great importance.

One of the men keenly interested in solar sailing was Dr. T. C. Tsu, an aerodynamicist at Westinghouse Research Laboratories. Dr. Tsu described the familiar technique of placing his proposed vehicle in orbit with conventional rocket power. Then, however, the spacecraft takes on a weird configuration. The gondola carrying crew and equipment unfurls a monstrous, parachute-shaped sail of plastic or aluminum foil 1,600 feet in diameter.

To send a one-ton payload to Mars by rocket would require more than eighty tons of propellant for the round trip. A thin plastic solar sail, even one 1,600 feet across, would weigh only one thousand pounds and tow a payload of equal weight. Added to this fuel advantage is the fact that a solar sail should be cheaper than any other power plant for space flight. It also requires no new knowledge, since light, high-strength plastics are common at low prices today.

Dr. Tsu's figures showed that while it would take us about 260 days to travel through space from Earth to Mars with chemical-rocket power, the sailing trip, despite a slow start, would take only 118 days! With the solar sail furled, or "tacked," to play sunlight against gravity, control in space would be possible, and a more nearly straight-line trip could be made. Dr. Tsu suggested, too, that the size of the sail would make it valuable as an antenna for long-range communication.

The most important possibility may well be the ultimate speed of the solar sailboat. While acceleration is obviously very slow, and a hot-rodder would die of boredom on a short trip, the speed constantly increases with time. The limit would seem to be the speed of light itself, in which case the sail would speed along on the crest of the light waves like a surfboarder who knows his business. Nautical skippers can sail their boats faster than the wind. It might be enlightening to see what happens to the theory that says light is the fastest speed possible, and that no other substance can achieve that 186,000-mile-a-second rate. Most scientists are conservative on this point, however, and cautiously mention speeds of "only" about 2,000 miles per second.

As we consider this 1,600-foot sail tugging our space capsule along in bright sunlight, we will realize that it is another potential source of power—electrical power. Schemes for using the photoelectric effect on thin plastic films have been proposed, and figures of "a kilowatt per kilogram" have been quoted as goals. Thus a solar sail might also develop many kilowatts for use in operating auxiliary equipment in the space vehicle, or for driving an ion-propulsion plant.

Getting Down to Earth

Out in the clear hard vacuum of space, solar energy at last came into its own for two reasons. There is more sunshine there than on Earth. And that's about *all* there is. A space-ship traveling to Venus could not stop at filling stations along the way, but perhaps solar energy would make such stops unnecessary. All of space is one vast solar fuel tank.

While many authorities feel that nuclear power will be the ultimate in space, a top official of NASA recently stated that *all* space vehicles to be launched for the next several years will use solar power of one kind or another. Eventually the applications may progress to thermoelectricity, thermionic conversion, thermal-mechanical engines using liquid metals as a working medium, and perhaps such more sophisticated systems as solar-heated hydrogen jet engines and ion propulsion.

There is a boom in solar energy fostered by man's leap into

space. The lowly sunbeam that couldn't make the grade as a cooker for Asians or house heater for Americans is at least a glamorous Space Age baby and growing on the diet of dollars from various governmental agencies eager for space power.

Wartime development of radar eventually helped bring TV into most homes. The use of solar energy as a space power source has already hinted at the possibility of global communication and education through such a medium, and may lead to many more practical applications on Earth. Man found helium in the sun; more recently he is discovering solar energy applications by looking into space.

It is not necessary to light a candle in the sun.

Algernon Sidney, "Discourses on Government"

6 PUTTING THE SUN TO WORK

In February 1972, NASA administrators at Houston announced that the major effort at that facility would be shifted from the moon back to the Earth, and that by 1973 perhaps three-fourths of its scientists and engineers would be concentrating on Earth-related problems. One important area would be that of remote sensing by the Earth Resources Technology Satellites toward better use of natural resources in farming, mining, and so on. Much to be desired would be a similar shifting of interest toward terrestrial applications for solar energy.

The backward nations could be a prime factor in the early use of solar energy. It is in remote areas that people are generally the poorest, both in money and energy. To build a fission power plant in an African village would be uneconomical; a fusion power plant would obviously be out of the question. To build either in a large city and transmit the energy through power lines over very great distances might be economically unsound. Solar energy, on the other hand, with its simplicity and low temperatures, lends itself readily to small installations capable of being tended by unskilled labor. Engineers have already shown that such plants could be built in some areas today more economically than fuel could be hauled in or power lines erected. Solar energy technology is here—ready to be used. It is not, like nuclear fusion, merely a gleam in the eye of a scientist. Even in the United States, where cheap power is abundantly available, solar energy is finding occasional opportunities.

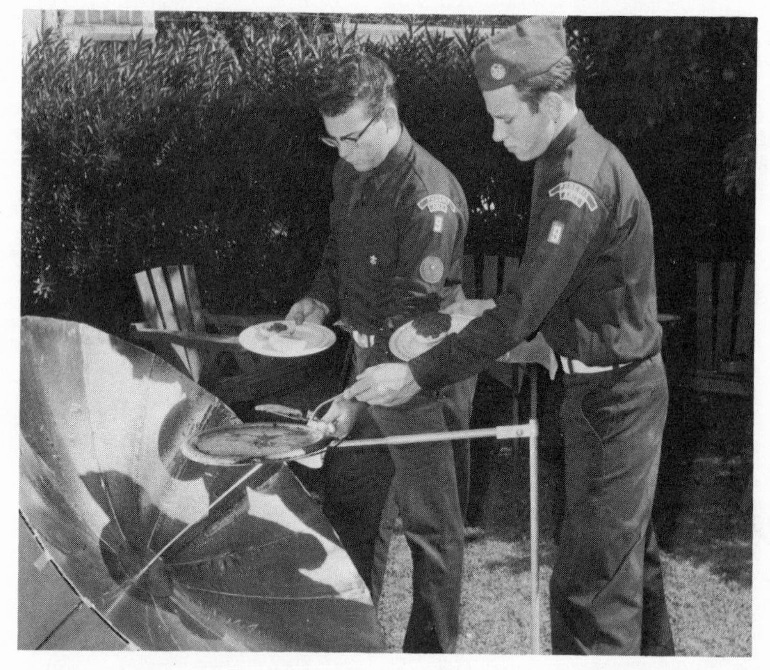

FIG. 42. Hamburgers à *la* solar energy.

The need for a new source of energy is imminent, and the potential of the sun to supply it has been demonstrated. What, then, are we doing about it besides talking and wishing? Certainly we have not done as much as we could, but a small dedicated group has made a token start. Let us discuss the uses to which solar energy can be put here on Earth.

Cooking with Sunshine

Of all the varied applications of solar energy one of the most striking—and useful—is at the same time the homeliest means of putting the sun to work. It is the sun that furnishes the heat for frying eggs on railroad tracks to fill newspaper space on otherwise dull days. Happily, there are better ways of cooking with sunshine.

We noted that solar pioneers cooked with solar heat more than a century ago. Rediscovered in the mid-1950s, solar cooking has proved itself effective, economical, and safe. Simple reflector cookers of a variety of materials ranging from cardboard and plastics to metal offer the equivalent of a 500-watt hot plate in bright sun.

For a time there were several types of reflector cookers on the market. Perhaps the most popular was the "Umbroiler," an umbrella-like reflector equipped with a grill. Designed by solar scientist George Löf, this cooker is easily packed in a box for transporting, can be set up in a minute or so, and in bright sunlight can boil a quart of water or broil hot dogs or hamburgers in about twenty minutes.

FIG. 43. Dr. Maria Telkes demonstrates her solar oven for colleagues.

Campers have not deluged the manufacturer with orders but the solar cooker is a practical piece of outdoor gear. It will broil steaks or cook bacon and eggs before a bed of charcoal can be burning well. For trips there is no need to take along fuel of any kind, which is a saving of money and space. No matches are needed, and there is no bothersome smoke. The solar stove is also far safer with respect to forest fires than are other means of cooking.

Akin to the reflector stove is the solar oven. Also portable, this unit is light in weight and some designs yield temperatures as high as 450 degrees F. Equipped with folding reflectors to increase the amount of heat trapped, the oven works on the "greenhouse" principle. Heat goes in through the glass top, but is not radiated back out. More compact and easy to handle than the reflector cooker, the oven need not be as accurately pointed toward the sun. Some ovens have been equipped with special containers of heat-storing material and can cook for several hours after sundown. The obvious disadvantage of solar cookers or ovens is their dependence on the sun. At night, in the shade, or during a rain, they are of little use; however, camp stoves are generally intended to be used during daylight, and in good weather.

With the great interest being shown in ecology and the preservation of resources and the environment, it is remarkable that so little use has been made of solar cookers and ovens by campers and for back-yard cooking. Here is an application that is both fun and an ideal way to conserve energy and reduce pollution.

The Industrious Sun

Where the solar cooker is impressive, the solar furnace is spectacular. In fact, a favorite trick of public relations men (though it must depress scientists) is that of shoving a two-by-four into the "hot spot," where it catches fire almost instantly. With a big enough furnace a steel beam can be substituted for wood and shortly turned into molten metal. These feats are like using a telescope to look into bedroom windows, but perhaps they are worth the scientists' anguish as attention getters.

Archimedes used the concentration principle to produce fire at a distance; the scientific or industrial solar furnace is focused closer at hand and is thus a more controllable tool. Modest-sized solar furnaces can be built by hobbyists, and these have been used for tasks ranging from the firing of ceramics to the brazing, soldering, and welding of high-temperature metals. Do-it-yourself furnaces deliver temperatures of about 2,000 degrees F., sufficient for a number of applications, including the burning of fingers. And even with these small furnaces it has been found wise to wear dark glasses, to protect against the white-hot glare of the focal point.

After World War II, with the upsurge in high-temperature research for fast-flying aircraft and missiles, scientists turned to the solar furnace to produce the elevated temperatures needed. There were plenty of them available, in the form of surplus searchlights. Searchlight mirrors are designed to reflect the light boiling out in all directions from the carbon arc in a tidy parallel beam. The parabolic shape of the mirror has this property, and of course it works both ways: parallel rays of the sun striking the mirror are reflected to the focal point of the mirror. The better the mirror the smaller the spot and the greater the temperature.

Even the aging searchlights did a commendable job, and many aircraft firms and others in the field mounted makeshift furnaces on the roof and worked out tracking rigs to follow the sun. The bright spot of light at the focal point might be several thousand degrees Fahrenheit, and all manner of materials were melted or oxidized. A few furnaces, some as large as ten feet in diameter, were specially made for the purpose, but for a really large solar furnace it was necessary to look to France.

At the end of World War II, scientist Dr. Felix Trombe interested the French government in building a solar furnace for research. There was an old fort at Mont Louis in the Pyrenees, high enough for the air to be very clear and favored with much sunshine. It was, Trombe felt, an ideal spot for solar research. The government agreed, and with amazingly modest funds Trombe and his colleagues built what was at that time the largest solar furnace in the world.

Spanning thirty-five feet, the huge vertical parabolic collector was made up of thousands of small mirrors accurately curved with tiny adjusting screws to focus at one point. Some distance away a large flat mirror called a "heliostat" was mounted so that it could move horizontally and vertically to follow the sun. Thus, during daylight hours the sun's radiation was reflected by the heliostat onto the parabolic mirror and thence to the focal point.

Developing seventy-five kilowatts of heat, the furnace was used for high-temperature research and also as an industrial smelter. Batches of up to a hundred pounds of material are smelted in the "pure" heat of the sun. There are no impurities in sunshine, so

FIG. 44. Dr. Trombe's large solar furnace generates one thousand kilowatts of thermal power.

there is no contamination as with other means of heating. Neither is there a magnetic field, as is often the case with electrical induction heating.

Trombe's idea paid big dividends in research done at Mont Louis. In the United States an estimated two hundred furnaces were in use at one survey, but for some time none of these approached the size of the Mont Louis furnace or even that of a twenty-five-foot mirror the French had built in Algeria to be used, among other things, to "fix" nitrogen out of the atmosphere for use as fertilizer.

Meanwhile, Trombe got busy on another solar furnace at Odeillo, France. Again using an array of adjustable mirrors, he built the parabolic reflector into the side of a huge laboratory building. The resulting collector was 115 feet high and 165 feet across. Instead of a heat output of 75 kilowatts, it has a maximum of 1,000 kilowatts. At the 1971 Solar Energy Society (the new name of the Association for Applied Solar Energy) Conference in Washington, D.C., Trombe and his colleagues reported on its use as a research tool for high-temperature materials, and also as an industrial smelting facility.

Solar Air-Conditioning

Thus far we have considered using solar heat for cooking and also for scientific and industrial furnaces. Important as these areas are, they are minor compared with another potential of solar energy—that of heating our houses and buildings.

It is one of the ironies of solar science that while each square yard of roof is receiving more than a horsepower of solar radiation when the sun is straight overhead, we do our best to bounce that heat off—and at the same time operate an expensive and fuel-consuming refrigeration unit to keep the house cool.

Worldwide, as we have said, about one-quarter of the fuel burned is to provide space heating. The developed countries have added cooling to that load, and as anyone who lives in a warm climate knows, the fuel bill in summer is higher than that in winter for heating.

Since we need heat for our houses, and the sun furnishes that commodity directly, here would seem to be a profitable application of solar energy to offset the problems of fuel depletion. We noted that a government survey in 1952 indicated a potential market of about 13 million solar-heated homes by 1975. Unfortunately, it will be difficult to find even thirteen in this country by that time. Not because it can't be done; it was done years ago. The problem is not one of technical complexity. In fact, the most successful designs have been the simplest.

Basically the problem is one of getting solar heat into the living quarters and keeping it there. This entails a heat collector of some sort, a storage medium, and a circulating system to move heated air or water. A variety of collectors have been designed and tested. Heat can be stored in water, rocks, or even in certain salts that store great quantities of heat. Overall, the most successful solar-heated homes seem to be those designed and built by inventor-engineer Harry Thomason of Washington, D.C.

In the Thomason design, part of the roof acts as a heat collector through which hundreds of gallons of water are circulated. Heat transferred to the water is stored in a tank in the basement, surrounded additionally by rock-heat storage. A small blower circulates the warm air. Ingeniously designed despite its simplicity and low price, this system heats the house in winter, heats water the year round, and even helps cool the house in summer. To do the latter, water is circulated from the storage tank to the roof, where it is chilled by evaporation during the night. This water is drawn on during the day to cool the house.

Because of the possibility of long periods with insufficient sunshine to heat the storage tank, the house is equipped with a standby heating system. Seemingly an admission of defeat, this is a more or less standard design philosophy that has evolved for solar space heating. While it is possible to provide all the heat necessary with solar collectors, it is not always economical to do so. In sunny

FIG. 45. Office building heated by solar energy in New Mexico.

climates at low latitudes a small collector and storage system will provide all the necessary heat. However, to do this in New England or Canada would require a huge collector and a huge storage area. There is thus a point of diminishing returns, and engineers find it wise to plan for only a percentage of the total heat requirement with solar energy. To carry over through long cloudy periods or continued very low temperatures, supplementary conventional fuels are provided.

Washington, D.C., is just at the edge of the "solar belt," almost at the 40th parallel of latitude, and would thus seem a marginal location for solar heating. Thomason's success is therefore very encouraging. To make the system even more efficient, he also heats a swimming pool with solar energy, and further bonuses are an

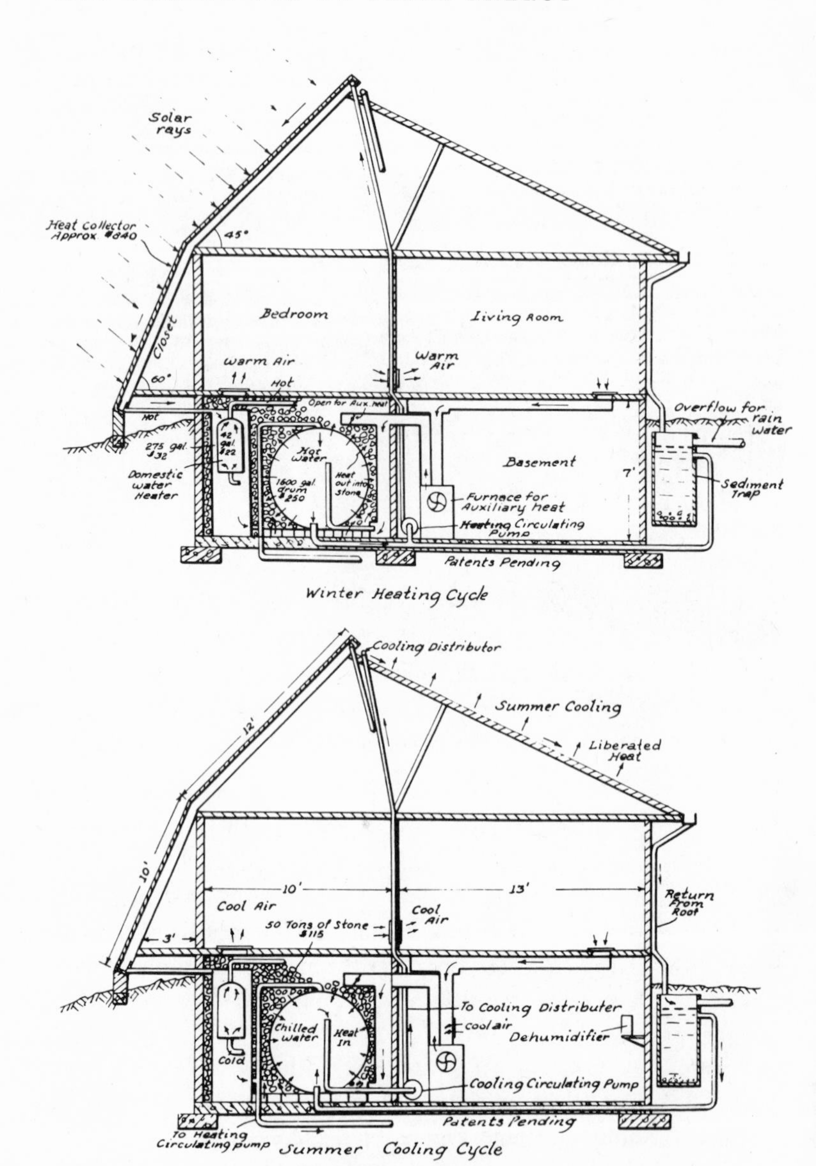

FIG. 46. Design of solar heating system designed and built into several houses in Washington, D.C., by Harry Thomason. The house is cooled in summer by circulating water to the roof, where it radiates heat to the night sky.

emergency water supply for domestic purposes or for fire-fighting.

The sun can refrigerate a home as well as heat it, and many experts think that cooling actually offers a greater potential for solar energy, since it is at its peak when needed most. This is just the reverse of the situation when sunshine is used for heat. Solar refrigeration systems have been built and their feasibility has been demonstrated; the economics remain a stumbling block. It is still cheaper to air-condition a house with electricity or gas. For example, one experimental solar cooling system included a copper roof that cost several thousand dollars! The answer lies in paring costs to make solar energy methods competitive. A breakthrough in materials, or perfection of existing designs, coupled with wide demand, could make solar air-conditioning practical in a few years.

Again, situations dictate developments. Just as solar panels at a quarter of a million dollars may be a bargain for a kilowatt of power a thousand miles out in space, a solar refrigerated and lighted home a thousand miles from the nearest utility might also be a bargain. While a gallon of oil may give us hundreds of thousands of Btus for just a quarter, if it costs us a dollar to haul in that gallon, the economy takes a decided plunge, and solar power at a capital cost that looks at first glance like a hopeless extravagance becomes a better buy.

While Thomason's solar homes are of very simple design, engineer Harold Hay has gone much further in this direction. His model house, built and tested in 1967 and 1968 in Phoenix, Arizona, used no conventional heating or cooling but relied on insulation and movable panels on the roof, which alternately covered and exposed shallow ponds of water. With air temperatures about 100 degrees F. in summer, the small building was kept below 77 degrees during dry months, and below 82 degrees in periods of high humidity. Hay reported that evaporation of less than a pound and a half of water produced the effect of one ton of conventional refrigeration. In winter the house temperature was maintained for the most part close to 70 degrees, with brief dips to as low as 65 during very cold weather.

FIG. 47. Simple solar-heated house, using movable panels on roof, and shallow tanks of water to store heat. The building is also kept cool without using mechanical equipment.

Here is solar air-conditioning at its basic minimum, and not sophisticated enough for many tastes. However, there are still many areas where such performance would be a vast improvement. The radiation cooling techniques used by Hay—not solar energy use but something like the exact opposite—have also been used by researchers in France to cool test buildings, and also to produce ice without mechanical equipment. Hay points out that centuries ago the residents of Teheran and Calcutta erected high "ice walls" and produced ice in their shade (even with air temperatures as high as 48 degrees F.) by radiating heat away to the clear night sky.

At least two solar homes have included swimming pools in the heating system, and hundreds of pools are being heated separately by homemade solar collectors. While a large collector area is needed and sometimes poses architectural problems, the attractiveness of the scheme is obvious when it is remembered that a gas heater costs several hundred dollars to buy and up to a hundred dollars a month to operate.

While solar energy does not yet heat a great number of homes, or even swimming pools, domestic solar water heaters abound in Florida. This success depends on several factors. First is the large amount of sunshine available. Improved designs, competitive costs of operation, and aggressive manufacturing and sales organizations add to the natural blessing of the state to put the sun in business.

The solar water heater has been with us for many years, and in some southern states aging pioneer installations can still be seen in operation. California in particular took to the solar heater and its rooftop glass-covered heating coil. But technical "progress" and financial affluence outmoded these early power savers. In present installations standby electric or gas fuel is called on when the tank temperature drops too low. In the meantime, appreciable fuel savings are realized.

Mountain cabins and remote camp sites are logical customers for solar water heaters; perhaps portable units may soon be available for campers.

Easing the Water Crisis

Along with many other shortages we also seem to be running out of fresh water for drinking and irrigation. Already the distillation of sea water has been resorted to in many locations in the world, and here again is a broad area of application for solar energy. When water evaporates, the impurities are not taken up in the vapor. Thus, when the vapor condenses on an appropriate surface it is pure and suitable for drinking, irrigation, industrial uses, and so on. Distilling water requires heat, the most easily obtainable form of solar energy.

The Office of Saline Water, Department of the Interior, has sponsored research in solar stills, along with its other research in the field of distillation. Battelle Memorial Institute is among those who have been active in this work, with a research station at Ponce de Leon Lighthouse Reservation near Daytona Beach, Florida. Franklin Institute, the Georgia Institute of Technology, New York University, Bjorksten Research Laboratories, Inc., Dr. George Löf, and New Mexico Highlands University have also done research for OSW. The Massachusetts Institute of Technology and the University of California have also worked on solar stills of various designs. The main problem is to get the cost of distillation per gallon down to a competitive price.

Cheap fuel is the solar still's competition. Conventional distillation plants produce fresh water for thirty cents per thousand gallons, and solar plants presently cannot match this figure. Initial costs are high, particularly when construction is of concrete and glass for long life. Saving money with plastic covers instead of glass has been accomplished, but hurricane Donna wrecked a promising air-supported plastic still in 1960. It is of interest that the still withstood the gale until power for the air pumps failed, however, and that far heavier structures were also wrecked by the storm.

Reporting at the Rome Energy Conference in 1961, Department of the Interior scientists gave estimates for a hundred-thousand-gallon-a-day solar still based on work with pilot plants. Such a still would cost $1 million and deliver fresh water at a cost of $1.22 per thousand gallons, several times that of conventional distillation methods.

Carl Hodges and his colleagues at the University of Arizona designed and built a solar still at Puerto Peñasco on the Gulf of California, to provide fresh water for a community that had up to that time been forced to truck water in at high cost. However, this still was redesigned from a solar-heated plant to one using the waste heat of a diesel engine employed in an experimental hydroponics structure. The engine pumped air to support the plastic enclosure, and produced carbon dioxide to enrich the pressurized atmosphere for enhanced growth of vegetables. It was found that the diesel could also furnish the heat for desalinating water for irrigating the hydroponic crop.

A number of other fairly large solar stills have been constructed.

FIG. 48. Solar still built in the town square on the island of Symi in the Mediterranean.

Some experimental designs were built in Australia (the headquarters for the Solar Energy Society is there) but the most impressive are those used on Greek islands critically short of water. In this Aegean archipelago, where Archimedes used solar energy as a military weapon centuries ago, the government embarked on an ambitious program of building nine sizable stills to eliminate the necessity of shipping in fresh water at great expense.

One of the first such projects converted the town square of Symi to a plastic still providing domestic water to augment the meager rainfall collected in cisterns. The largest of the nine stills is that built on the Isle of Patmos. Looking remarkably like the tilted glass still built a century ago in Chile, the Patmos design uses an aluminum framework, and a base of rubber sheeting to hold the salt water. Covering an area of 130,000 square feet, the still produces about seven thousand gallons of fresh water a day. Additionally, it collects rainwater and adds this to the reservoirs. Thus each

FIG. 49. The Japanese install "earth-water" solar stills on a dry volcanic island.

of the two thousand residents of Patmos receives several gallons of fresh water a day thanks to the sun.

Solar stills are being built on a very modest scale in the United States by the Sunwater Company of San Diego. Designed by Horace McCracken, the stills are constructed of concrete in modular units three feet by ten feet. McCracken reported in 1971 that his firm had sold seventy-five installations. Buyers are generally in remote areas where fresh water is in short supply, and McCracken foresees the establishment of "village industries" turning out increasingly large areas of stills.

As the water crisis grows, and as fuel costs continue to climb, the attractiveness of solar distillation may offset the disadvantages of capital costs and large land area needed. As an example, studies of desalination methods for the Los Angeles area disclosed the disconcerting fact that while it would be straightforward engineering to build a gas-fired plant capable of providing ample water from the ocean, such a plant would require *all* the natural gas available.

A Solar "Earth-Water" Still

One remarkable result of solar still research is the "earth-water" still, invented by a Japanese named Kobyashi during World War II for production of water on the dry islands of the Pacific. Little came of the idea at that time, but in recent years Japan has produced a variety of small glass and metal stills capable of producing fresh water—not from salt water, but from the ground!

Two Department of Agriculture scientists, Ray Jackson and Cornelius van Bavel, independently developed a similar though much simpler survival still for use in the desert. A cone-shaped hole is dug in the ground and a container placed at the center of the depression. Next a sheet of clear plastic is laid over the hole and the edges covered with earth. The plastic is pushed down in the center and weighted with a rock, forming a crude plastic still with sloping sides terminating just above the collector. Heat generated under the plastic evaporates moisture in the earth, and this condenses on the underside of the plastic and runs down to drip into the collector. More moisture is drawn from deeper in the soil by capillary action. Properly set up, the earth-water still will produce as much as a quart of water a day from about a square yard of earth.

While the earth-water still is a very specialized use of solar energy and will make no dent in fuel consumption, some fliers make it a practice to carry such a survival kit, and one life has been reportedly saved in this way. There may also be potential uses in agriculture, swamp drainage, and so on.

Using Solar Electricity

Moving from the uses of solar heat we come to applications involving solar-generated electricity. Most of us are more familiar with what solar batteries are doing in space than we are with their accomplishments here on Earth. While the total kilowatt-hours produced by such terrestrial applications do not add up to enough to worry utility companies, there are a few devices now drawing token electric power from the sun.

Solar-powered radios in a variety of shapes and sizes have been manufactured, from tiny transistor models using one solar battery to long-range communication equipment powered by large panels.

FIG. 50. Baker electric car, powered by the large solar panel mounted on the roof.

Solar batteries have been used in Africa and South America, in experiments with "listening stations," and solar-powered emergency signaling stations along highways and freeways have been tested. Since such units are self-contained and require no power lines and little maintenance, they are economically attractive even now. In addition, phonographs have been run on sunlight, as have clocks, electric shavers, sewing machines, flashlights, and beacons in lighthouses and airways markers.

The "Solar King" panel built by International Rectifier Corporation was used to power a Baker electric car. Mounted on the roof of the antique 1912 auto, this large solar panel charged the storage batteries in the car, and after sufficient exposure to sunlight the car could be driven for several miles, actually on sun power. The project was, of course, a demonstration, and there has been no rash of solar-powered electrics blocking traffic on the freeways.

The solar-powered boat built by the Agency for International Development for use in South America, while not as racy looking as the conventional marine craft, actually was powered by the sun, and the fact that it ended in a controversy does not shoot down the concept. An educated guess as to the future of transportation is that it will be electrically powered—with electricity possibly generated by solar energy, though not necessarily on the roof of the vehicle itself. Perhaps the car will be plugged into a wall socket in the garage to charge up from a collector on the housetop.

The Sun Roof Power Plant

Use of the rooftop to gather heat for warming or cooling a home is a not-too-far-fetched idea. A trip to the attic in summer indicates the amount of heat that finds its way even through roofing material designed to keep it out. But the roof as a power plant for generation of electricity for lighting, cooking, communication, entertainment, and miscellaneous power needs for mowing, workshop, and so on, is a more challenging idea.

The heat falling on a home is sufficient to warm it in winter and cool it in summer. Beyond that, it is enough to provide all the other energy needs of the house, as described above, with a good margin of safety—if we can learn ways of availing ourselves of it. Back in the early days of the solar battery, the solar-shingled roof was an eye-catching idea. With a roof twenty feet by forty feet, converting solar energy to electricity at 10 percent efficiency, only five days of sunshine a month would provide sufficient kilowatt-hours to service a home. The obvious fly in the ointment then, as now, was the price of the "solar shingles." At current prices such a roof would still cost hundreds of thousands of dollars and would take many lifetimes to pay off even with no maintenance or replacement costs.

The solar roof became the butt of ridicule, and its proponents were accused of overselling solar energy. More recently, however, the idea has moved from advertising brochure to scientific journal, and learned treatises have been written by scientists and engineers proposing these rooftop power plants. One writer even suggested tying in the home power plant with the public utility. When the roof received an excess of power, this surplus would be carried off by wires for use elsewhere. The watt-hour meter would in effect run backward at these times to lower the householder's bill. Appealing as this arrangement might be to those of us paying the electric bills, it has been deemed uneconomical by other experts in the power field. Why suffer the losses in the transmission lines? they argue. Why not build storage facilities into each house and have a self-contained power plant on each roof? No longer would we need unsightly poles and wires that are also a hazard and an expense. And electric power would be no more costly in rural or even remote areas than in town.

Such a mass-production development will depend on one of two things. Either the public must demand solar power, or industry and government must take the initiative to produce solar power converters on a speculative basis. Neither of these seems about to happen at the moment. Many proponents of applied solar energy feel that a safer approach to more exploitation of the sun will be the building of large solar power plants to supplement existing fossil-fueled and nuclear plants. Such possibilities will be discussed in later chapters.

The sun has demonstrated that it can heat our homes and cool them. It can cook food and heat water, distilling the water first if need be. Solar heat is available at high temperatures and at a purity unmatched by other sources. Solar electricity is here, although not yet at a penny a kilowatt-hour.

What little use man has made of solar energy in his machines, heaters, and the like has employed simple physical or mechanical laws. There is another side to the radiation from the sun, the chemical side that is much harder to fathom and even harder to duplicate. In the next chapter we will look at the processes of photochemistry.

And God said, Let there be light; and there was light.

Gen. 1:3

7 THE WHITE MAGIC OF PHOTOCHEMISTRY

The stored-up fossil fuels man has inherited on Earth, the coal and oil that nature has stockpiled slowly through eons of time, are products of a single phenomenon, that of photosynthesis. Logically, then, it would seem that as man began to convert solar energy into power he would look to nature in an effort to copy this method which produced all his fossil fuels and also the food that keeps him alive.

Strangely, man has done least in harnessing sun power with the method used so successfully by nature. We have talked of electrical or mechanical conversion and pointed out that these have been developed to a fair degree of efficiency. These are physical methods that find no counterpart in nature, surely not in the phenomena we will concern ourselves with in this chapter: those of photochemistry.

Photochemistry is the study of chemical reactions produced by light, including the near ultraviolet, the visible, and the near infrared. There are two common photochemical reactions man profits from: photography and photosynthesis. In photography man has done something nature does not do; in photosynthesis the opposite is the case. Man would dearly love to duplicate the process of photosynthesis, and does not only because thus far he cannot.

Photosynthesis: Sunlight into Fuel

While the definition of photochemistry is simple, the process itself defies man's detective work to analyze it. It is a highly

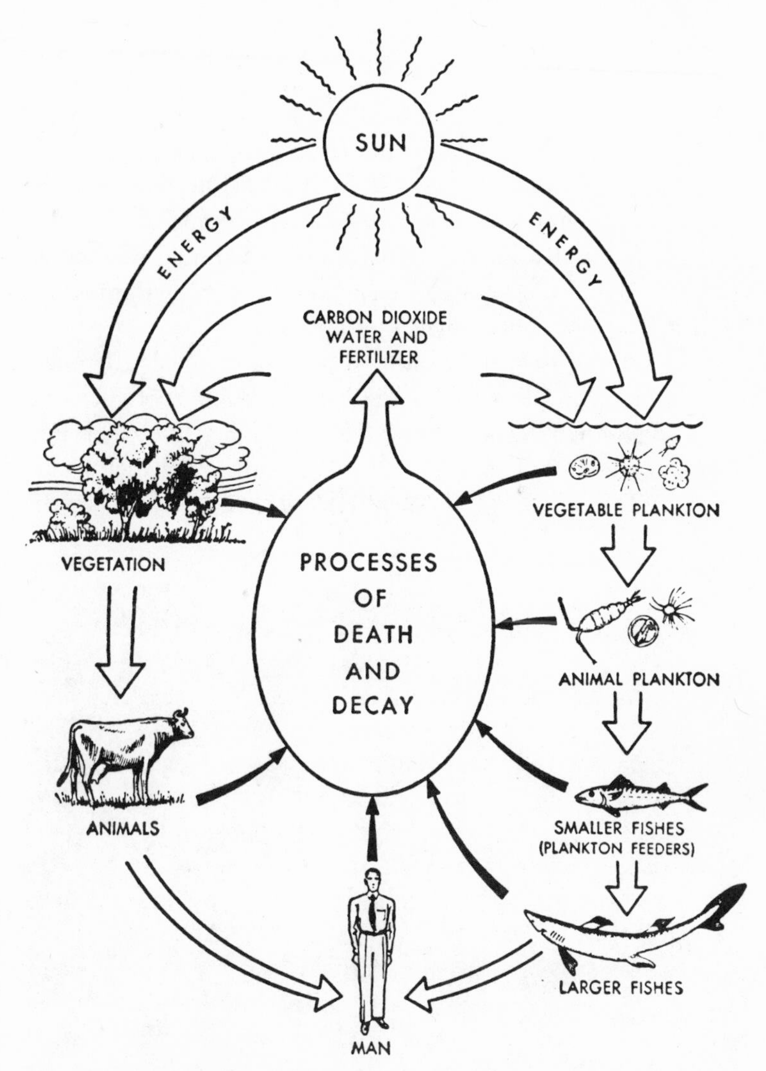

FIG. 51. The solar energy cycle in nature.

complex happening which man can describe, in terms including photons, Planck's constant, and even "einsteins'" of light, but which he cannot explain.

Plant life is a living example of the conversion of solar energy to stored chemical energy. A wheat field soaks up the sunshine bathing it and changes the light into calories of heat energy in the plant. This energy is stored for months, or even years in huge silos, is finally made into flour, baked into bread, and eaten to power our bodies for their various tasks. With the apathy accorded the commonplace, most of us look with little wonder and scant respect on the growing plant, unaware that it is sophisticated beyond our scientific comprehension. The conversion of sunlight into electricity in a solar battery is much more exciting.

The great part that photosynthesis plays in life on Earth comes into better focus when we consider the magnitude of the work plants do each year. It is no mean feat, combining 100 billion tons of carbon and 25 billion tons of hydrogen into energy-rich hydrocarbons, and producing 100 billion tons of oxygen in the process.

FIG. 52. Improving the natural photosynthetic process with plastic mulch.

If these busy green plants were to disappear, so, too, would life as we know it. Man needs heat energy to live and is constantly bathed in such energy from the sun. But he cannot convert sunlight into food himself. Man in space could not exist without stored-up food; he can on Earth because plants turn sunlight into a source of energy compatible with his body structure.

Many authorities, including Dr. Farrington Daniels, feel that photochemistry, of which the photosynthesis of plants is nature's prime example, is our ultimate hope in putting the wasted energy of the sun to practical use.

Photography, an aspect of photochemistry that man has more or less mastered, has a long history of study but not as long as that of photosynthesis. Even primitive man was aware that the sun made his crops grow, although for thousands of years he was content merely to accept that fact and make an occasional sacrifice to the sun-god. But as scientific questioning supplanted cult and magic, men began to study the relationship of sunlight and plants. During the seventeenth century, a Flemish chemist named Van Helmont conducted the first scientific studies of plant growth, and made the astonishing discovery that the constituents did not come from the ground as had been thought. Instead, they must come directly from water, air, and from the sun.

In our century scientists have begun to pin down the "quantum" efficiency of photochemistry and to dig into the riddle of how chlorophyll changes light into chemical energy. But even with nearly three hundred years of investigation, and a large body of formal literature, little is known today of how the plant achieves this deceptively simple transformation of energy from light to chemical storage.

Essentially, photosynthesis consists of the reduction of carbon dioxide to form a sugar, and the splitting of water to produce oxygen. One writer suggests the analogy of a hand-cranked mill, into the top of which is dumped water and carbon dioxide. The crank is turned, its energy analogous to that of sunshine, and out of the bottom come sugar and oxygen. This marvelous machine, unfortunately, is something like the goose that lays the golden eggs, for when we cut into it to see how the trick is done we ruin the wonderful mechanism and learn little for our pains. Thus the scientist has had to treat photosynthesis largely like the "black box" of his electronic brethren, in which only input and output are known.

We can, of course, rationalize this lack of progress in duplicating artificially the photosynthesis of nature. Although the growth of a plant is miraculous, it is also very low in efficiency. Corn is perhaps the best converter of sunshine, but it does that at an efficiency of only about 0.3 percent. Only because of the sun's effectiveness in producing great amounts of energy is there enough food for all the billions on earth. Some plants like wheat are less than half as efficient as corn, and the average is perhaps between 0.1 percent and 0.2 percent. So, for all the magic of chlorophyll, man might well wonder if the secret is worthwhile with a return of such low magnitude. The solar battery, remember, converts as much as 14 percent of total solar energy, a whopping percentage when compared with corn's tiny 0.3 percent. However, biologists, chemists, and, more recently, solar scientists are eagerly studying photosynthesis because the potential rewards are great.

Improving on Nature

The fanciful tale of Jack and the Beanstalk has an appeal, not just for youngsters, but also for those concerned about feeding the already hungry population of a rapidly growing world. The idea of dropping a seed into the soil and jumping back to avoid being lifted heavenward seems to have little chance of ever being more than just a dream, but there are some interesting developments to ponder.

Normal crop production is unlike Jack's magical beans, and is about three tons per acre per year. Despite tales of plants flourishing to music or well-intended prayers, the main thing that makes them grow is sunlight, combined, of course, with water and carbon dioxide. This is indicated in laboratory experiments producing freak plants that shoot up at a great rate because they are force-fed.

The German biochemist Otto Warburg claimed to have achieved efficiencies of 75 percent for photosynthetic processes carried out in the laboratory. Most American researchers feel that a more modest 30 percent is closer to the maximum realizable value. Even this figure is 150 times that of nature, and if it could be accomplished in the field would produce crops of 300 to 450 tons per acre per year, rather than the present yield of 2 or 3 tons per acre.

Such a yield would make Jack's beanstalk seem a lazy pot plant, and throw agriculture experts into a panic. Of course no such bumper crop as this is in sight, but when we recall that plants convert only about a thousandth of the energy available, the figure does not seem so unreasonable.

The Japanese have already pioneered growth boosters for plants like the gibberellins they isolated a number of years ago that lead to Jack-and-the-Beanstalk-like production in some plants. Dr. Farrington Daniels feels that through selection and proper plant breeding it should be possible to develop algae and plants that will give a much higher photosynthetic efficiency in concentrated sunlight and an atmosphere much richer in carbon dioxide. In fact, scientists have already begun work in the direction of producing more food per acre by giving nature a helping hand. One object of this photochemistry research is the algae *Chlorella pyrenoidosa*.

Algae Culture

In 1951 the Carnegie Institution engaged the firm of Arthur D. Little, Inc., to demonstrate the production of algae on a commercial scale. Work had earlier been done with laboratory growth of these microscopic organisms, but this was the first attempt at building what was in effect a pilot plant.

Algae are more familiar to most of us as seaweed, or the green scum growing on ponds. Their use as food is not particularly appealing at first glance, but then the tomato was considered poisonous for many years before it was first tried and accepted as a food. The plant that Arthur D. Little built in Boston was a spotless installation, and produced its algae not in open air on a scummy pond but in the clinical sterility of a closed environment.

A large plastic tank, laid out in an oval, covered six hundred square feet. In it twelve hundred gallons of water were circulated by a centrifugal pump and cooled by heat exchangers, since the algae refused to grow above certain temperatures. As the microorganisms circled their big racetrack, air containing 5 percent carbon dioxide was fed into the plastic tube. Thus sunlight was gotten to the algae more efficiently, and far more carbon dioxide than nature's provision of 0.3 percent was available.

The plant operated for about three months and in that time produced some seventy-five pounds of dry algae. For one acre, this converts to twenty tons per year compared with about three for dry land farming of the conventional type. Man, despite his inability to understand how photosynthesis takes place, was able to better nature appreciably in the first quantitative test of algae culture. Based on pilot work, a hundred-acre plant was next studied on paper. A yield of thirty-five tons per acre was estimated because of improvements to be made, and output for the plant would be more than six tons a day at a cost of about twenty-five cents per pound.

The Japanese have gone further with actual human consumption of algae, and compare them with other powdered products like green tea and seaweed. Since they have a strong aroma and taste of fish, algae are best mixed with something that goes well with that flavor. Noodles and rice crackers are suggestions, though the dark-green color of the algae darkens the food and may not be acceptable. The powder has been successfully added to candy, cake, and soft drinks.

A soybean soup called *miso* is a staple breakfast item with the

FIG. 53. Production of algae in experiments at the University of Florence conducted by the Italian Council for Research.

Japanese, and algae have been mixed about one part in eight with this soup. Ice creams and fish flakes were also tried. Algae seem to have about the nutritive value of yeast, but even less taste appeal. The Russians, in fact, object to them as a food for human beings except indirectly, as when they are fed to animals raised for meat.

More recent work with algae does not seem to have lowered

prices, and some researchers feel that fifty cents a pound is a more reasonable estimate of product cost in the marketplace. They state that this price will not be economically justified until it can be shown that there is special nutritive or other value to make algae more than a simple food substitute. Increasingly, algae are now being grown with deuterium, or heavy water, waste from atomic plants substituted for hydrogen. The resulting solids may well be of the "special value" to make the algae phase of photosynthetic research pay off.

Proponents of algae culture still claim they are on the right track, pointing out that there is room for much improvement in the annual yield—perhaps to a figure as high as 100 to 110 tons—and also that methods may be made more economical, as for example, by using strains of thermophilic, or heat tolerant, algae, so that cooling of the water will not be necessary. Beyond this is the undeniable truth that "artificial photosynthesis" like the algae culture that has been achieved is quite possible on land otherwise unusable for agriculture. It is also likely that when we run short of food we will be more than willing to pay a premium for methods that will produce ten times the normal yield. In the meantime, there are more immediate developments in the enhancement of nature's photosynthetic process.

Plastic Farming

On the horizon is the farm "sealed in plastic." Such controlled environments, with added CO_2 in the air (perhaps pressurized air at that), an optimum temperature maintained, nutrients added, and with little water lost in the process, are well past the laboratory stage. Until recently the plastic farm has not amounted to much more than an experimental or hobby technique; now there are a number of seemingly promising installations in various parts of the world.

The University of Arizona received a contract for a full-scale closed-environment agriculture system for the tiny sheikdom of

Abu Dhabi on the Persian Gulf. This will provide food and fresh water, both of which are in short supply locally in Abu Dhabi. The United Nations has also taken an interest in closed-environment farming, and is said to be providing funds for a pilot installation of plastic farming engineered by Hydroculture, Inc., of Phoenix, Arizona. The U.N. testing will be done in Lebanon. Meanwhile, Hydroculture has an apparently successful commercial venture in operation near Phoenix, and markets premium-quality tomatoes under the trade name of Magic Garden.

Steel-framed and covered with plastic, the hydroponic plots have concrete floors on which gravel serves as the plant bed. Water and nutrients are automatically fed to the plants, which grow eight feet tall on vertical cords. Temperature is controlled to 85 degrees F. during the day and 65 degrees F. at night. According to spokesmen, each mature tomato plant produces thirty pounds of market-

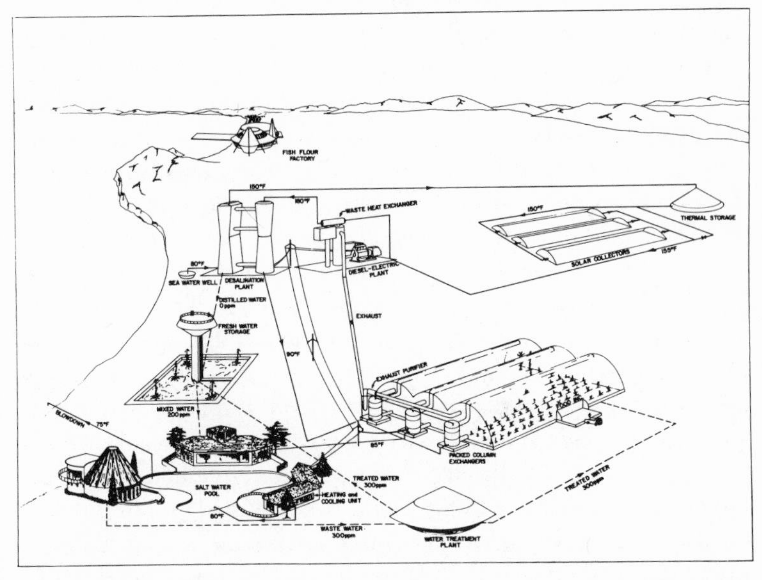

FIG. 54. System for production of power, fresh water, and food for a coastal desert community. Designed by Carl Hodges and others at the University of Arizona.

able tomatoes a year, compared with eight to ten pounds on conventional plants. Eight 26-by-128-foot greenhouses produce as much as from 4 to 8 acres of farmland.

Cucumbers and chard are also grown, and in another chamber, called the Magic Meadow, grass is cultivated for feeding livestock! One pound of oat or barley seed produces seven to nine pounds of grass eight to ten inches high in seven days.

Grow Your Own Fuel

The idea of using photosynthesis to produce power—in effect "short-term fossil fuels"—is intriguing, even though it seems doomed by arithmetic. Food may of course be considered a fuel, since it provides the motive power for our bodies. Gasoline produces calories in the internal combustion engine it is intended for; a stack of hot cakes does the same thing for the human machine. The calories we heavyweights count so worriedly are essentially the same as those that drive a steam engine; they are heat energy. There is a big difference, however, in nature's production of the hydrocarbon fuels we use and the carbohydrates we eat. Sufficient food is grown currently and continuously; the bulk of our fuels have been produced over millions of years, and only those like wood are of the "income" variety.

There are ways of converting woody products to liquid or gas fuels; for example, the Fischer-Tropsch process. This might take the thirty-five tons of algae grown on an acre and reduce it to from six to nine tons of fuel. Unfortunately this pleasant-sounding solution is more naïve than it sounds; a forest of the fastest growing trees known and covering the entire world would not do the trick. Even the more efficient algae culture method of converting sunlight to fuel would not work. If we burned all our crops to produce fuel, as has been done on a small scale during wars, we would still be short of fuel, to say nothing of hungry for food.

Quite recently, however, there has been a new development that may somewhat change the outlook for algae culture. Scientists have produced what they call "biopower," electricity from living cells. Such an application might lift photosynthesis into a possibility as a power source.

The Solar "Bug Battery"

We are used to the notion of living things turning sunlight into matter and accept this phenomenon on faith, though we don't know quite how it comes about. Nature's creatures also convert sunlight to electricity, generally in an indirect enough way for us to miss the significance. Galvani and his experiments with frog's legs proved the existence of electric current in living things. Most of us know that our brains operate with a feeble electric current, but the idea of producing appreciable amounts of power with living things is rather new.

In 1961 Dr. Frederick Sisler of the United States Geological Survey gave a demonstration of biochemical fuel cells in which decomposing organic material from the ocean bottom produced electricity. The test biocells converted sea water to electricity—two volts of it. Sisler also proposed using algae to convert sunlight into electric power.

Although little of a practical nature has come of the bug battery, proponents of biopower feel that eventually such "power plants" may produce electricity at a cost of only one mill per kilowatt-hour, making it competitive with nuclear and even conventional power plants. With a potential efficiency far greater than that of heat engines operating today, the bug battery is one of the intriguing possibilities on the horizon for solar energy.

There are less sophisticated methods of producing electricity from solar energy through biological converters. C. G. Golueke and W. J. Oswald some time ago generated methane gas from the fermentation of algae. The algae were grown in municipal sewage effluent, and results led the researchers to speculate on producing electricity for between one and two cents per kilowatt-hour by using the methane as fuel for a conventional power plant. A total efficiency of about 2 percent from solar energy to electricity was reported—not earthshaking, but far better than 0.3 percent, as achieved in natural processes. Estimates were that a one-acre algae pond could produce as much as fifteen kilowatts of electricity.

Chemical Storage of Solar Energy

One of the biggest problems of utilizing solar energy as a power source is that of storing energy for later use. An example is the problem of saving back sunlight for use during the darkness. To do this with solar batteries, or with any dynamic converter such as we considered in Chapter 5, involves adding storage batteries to the power system; this is costly and leads to further losses of efficiency. There is a great beauty in ideas that use photochemistry to convert the transient energy in sunlight into a stable and storable source of energy. Here would be not just a solar battery, but a solar storage battery.

Photosynthesis is only one example of the conversion and storage of solar energy. Fortunately there are other, simpler photochemical reactions. Scientists list eight primary processes: formation of free radicals, electron transfer, intramolecular rearrangement, photoisomerization, photoionization, photoconduction in solids, photosensitized decomposition of unexcited molecules, and photophysical processes such as fluorescence, phosphorescence, and so on.

The two processes generally considered most likely to succeed in the economical conversion of solar energy are the formation of free radicals and the transfer of electrons. In 1937 a breakthrough was made in photochemistry when the English biochemist Robin Hill found that the chloroplasts in green plant cells evolve oxygen from water when illuminated in the presence of certain ferric compounds. In the "Hill reaction," hydrogen and oxygen are separated when water is placed under ultraviolet light. Krasnovsky in Russia achieved a similar reaction with chlorophyll extracted from the chloroplasts.

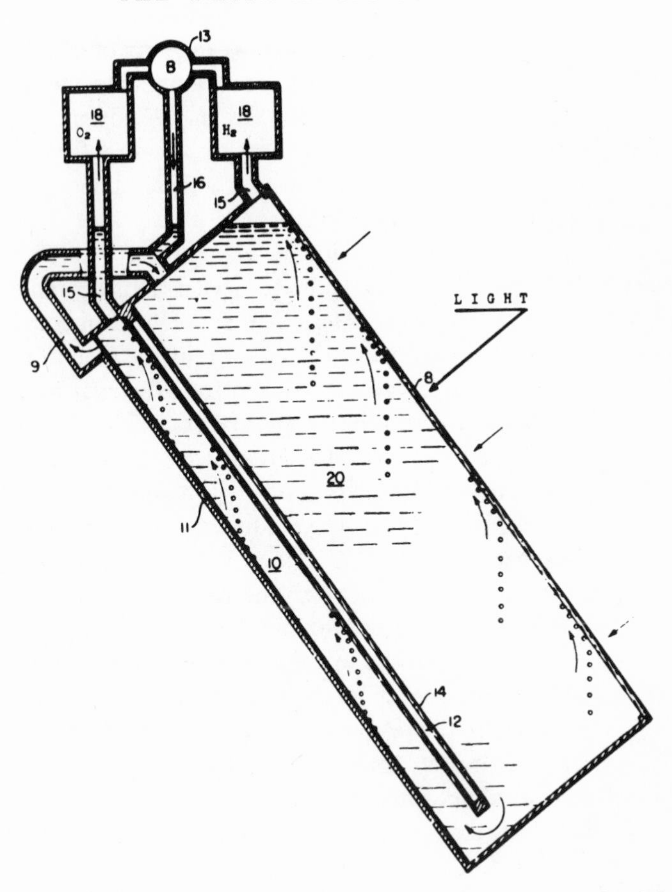

FIG. 55. The Hill reaction, in which sunlight is used to effect the breakdown of water into its constituent gases.

Other researchers, including Eugene Rabinowitch and Lawrence Heidt of the Massachusetts Institute of Technology, have demonstrated photochemical reactions of various types. It has been shown that ferrous solutions when illuminated acquire an electrical potential between the illuminated liquid and the dark portion. This is in effect a battery that converts light to electricity somewhat like the solar battery.

Heidt worked with solutions that had the property of decomposing water into hydrogen and oxygen when illuminated by ultraviolet light. While the reaction was accomplished only on a laboratory scale, it might be possible to couple it with a fuel cell to produce electricity. The hydrogen and oxygen recombine, and the water can be reused for further photochemical conversion.

This may seem merely an interesting curiosity until it is remembered that both oxygen and hydrogen are fuels, and that both are stable and storable. Here was a potential method of converting sunlight, not into electricity that must be used immediately or stored in a battery, but into gaseous fuels that can be easily handled and stored or transported over great distances.

One type of fuel cell is powered by oxygen and hydrogen, and a solar photolysis fuel-cell closed system was proposed as a power plant for space flight. In effect, it would be fueled by water used over and over again.

Although photochemical reaction thus far has resulted in the forming of only small amounts of gas, the feat has nevertheless been demonstrated. Other laboratory work suggests a conversion efficiency for some photochemical processes of 60 percent and higher, compared with the fraction of a percent found in nature.

Two Boeing Company engineers proposed another type of photochemical solar converter. Called a "solar concentration cell," the device would make use of certain phenomena occurring under ultraviolet radiation. The cell would have acid in each of its two sides, and light striking one would create an electrical potential between the two sides. This electricity could be used as generated or stored for later use. Potential advantages for the concentration cell are its low cost as compared with silicon solar batteries, and the elimination of heavy storage batteries. Further, it has a potential efficiency of 100 percent.

The Magic of Photochemistry

Photochemical conversion of solar energy is considered by some scientists to be the greatest hope for man in the future. By nearly all it is conceded to be one of the most difficult problems. Fortunately, many developments are working to the advantage of the photochemical process. Future space flight may require systems that photochemistry uniquely fits. Photosynthesis, for example, could make possible planetary voyages of several years by providing not only food and power but also water, fresh air, and the necessary waste disposal. A photochemical system that separates water into its constituent gases in sunlight and recombines them in darkness to produce electricity is ideal for orbital flight in which part of the time is spent in darkness.

Over the long haul, we must look to photochemistry for the photosynthesis of food, and perhaps fuel, to support our evergrowing and increasingly hungry population. But only a small amount of money has been spent on photochemical research for the promising results obtained.

Projections of the U.S. electrical power demands over the next 30 years indicate that the U.S. could be in grave danger from power shortages, undesirable effluence and thermal pollution. A pollution free method of converting solar energy directly into electrical power using photovoltaics on the ground shows that sunlight falling on about 1% of the land area of the 48 states could provide the total electrical power requirements of the U.S. in the year 1990. By utilizing and further developing some NASA technology, a new source of electrical power will become available. Such a development is attractive from conservation, social, ecological, economic and political standpoints.

William R. Cherry, NASA (1971)

8 ELECTRIC POWER FROM THE SUN

While there may be great potential for photochemical conversion of solar energy to a form suitable for power use on a large scale, such developments are presently in the realm of speculation. There are a number of other methods much closer to feasibility, and it is probably to one of these that we will be looking for supplemental power in the near future.

Most of the power we use is in the form of electricity, a very convenient and safe form of power. Consumption of electric power is increasing about twice as fast as are other forms, and it is expected that ultimately nearly all our requirements will be for electricity. Presently we produce electricity in a very roundabout way, using fuel of one kind or another to produce heat, which in turn drives mechanical engines, which then actuate generators to produce the electricity itself. It is of course possible to produce electricity in much more straightforward fashion, and we shall discuss this "direct conversion" method first.

FIG. 56. Solar batteries used to provide a perpetual power source in a remote, unattended location.

Solar Batteries: Light-Power

In a properly made solar battery exposed to light, electrons flow from one layer of the device to the other. All that is necessary is to attach conducting wires to the layers and we have a source of power that will play a radio, drive a motor, or charge a storage battery. Early solar batteries were round, the shape of wafers sawed from the ingot. Today most are rectangular, standardized to dimensions of 1 by 2 centimeters, and 40 millimeters thick. A single cell weighs about two grams and develops an output in milliwatts at approximately half a volt. Just as flashlight batteries are put together to give more power, solar batteries can be connected in series to raise the voltage, or in parallel to yield more current. "Shingled" modules of five cells give about a two-volt rating; twenty-eight-volt panels are fairly common in space power applications.

Other materials than silicon have been used. Gallium arsenide was one of the first; however, it is even more costly than silicon, and seems to have application only in high-temperature scientific projects. A more promising material is cadmium sulfide, and solar cells with efficiencies approaching 5 percent have been made with this material. The principle of operation of a solar battery made from cadmium sulfide is different from one made of silicon. Also, the material used in some experiments is transparent and can intercept light from all directions.

It is particularly interesting that fairly large cells, up to three inches square, have been produced. Obviously a major part of the cost of a sizable solar panel is in the assembly of hundreds or thousands of carefully matched cells. So workers with cadmium sulfide and other materials are looking for techniques that will permit production of large-area cells.

Proceeding from the inorganic materials, researchers are also doing work with solar batteries of an organic nature, as we have noted. Various dyes have been used with some success, and the potential here is a low-efficiency but very low-cost battery, producible in large sizes. Optimism prevails, and some researchers are talking in terms of one dollar to ten dollars per square foot of surface. When that day comes, we may well roof our homes with solar shingles.

FIG. 57. Large thermoelectric converter built by General Electric at its facility in Phoenix, Arizona.

The Hot Idea of Thermoelectricity

While the solar battery is the simplest, most direct conversion route from sunlight to electric power, it is not the only direct conversion method. Thermoelectric conversion changes the sun's heat into electricity. The discovery of thermoelectric conversion dates back to the "Seebeck effect" of 1822, when the German physicist experimented with heat and magnetism in various materials. Heat applied to a junction between two metals produces electricity. Furnace controls are an excellent example. A thermostat to regulate a gas furnace needs a current of electricity, originally supplied by conventional means from house wiring. The idea of inserting a "thermocouple" into the pilot flame on the burner was developed, and thermoelectricity generated sufficient current to operate the controls without the necessity of connecting to the house wiring and adding a transformer.

The recent flowering of thermoelectricity, like the solar battery, owes itself to the development of semiconductor materials. Pioneer attempts at thermocouples yielded efficiencies of 1 percent or less. Today materials like lead telluride and zinc antimony produce power at 10 percent efficiency and theoretically can do much better. What is needed in the thermocouple is a high "Seebeck coefficient," or ratio of power generation to electrical resistivity and thermal conductivity.

With the solar battery such a simple device, why spend time and money on developing thermoelectric converters that seem far more complicated? For instance, while solar batteries perform excellently without any kind of concentrators, the thermocouple is generally useless in the unfocused rays of the sun. The problems of waste heat and maintaining sufficient temperature difference between the hot and cold ends of the thermocouple further complicate the picture. Nevertheless, there is good reason for all the effort expended. A square yard of solar batteries costs in the neighborhood of fifteen thousand dollars, exclusive of the cost of assembling them into a usable panel. In the thermoelectric converter, a solar collector focuses this much sunlight onto thermocouples only a few inches in diameter, resulting in a great saving of expensive converter material. Since not as many wiring connections are needed, the reliability of the system also increases.

Present work with thermoelectricity began in Russia, even before the boost given by the developing semiconductor technology. In the early 1930s Dr. Abram Ioffe pioneered the field, predicting an efficiency of 4 percent for materials then available. The results were liquid-fueled thermoelectric generators powering radios and other electrical equipment in outlying areas. Thermoelectric refrigerators were produced commercially in 1953, and in 1956 Ioffe developed solar-powered forty- and one-hundred-watt thermoelectric converters.

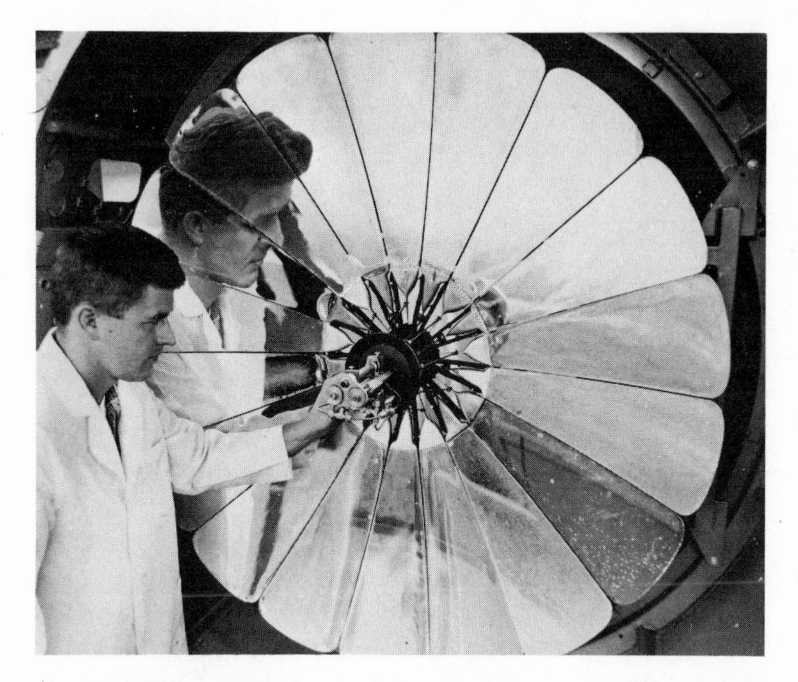

FIG. 58. Folding thermionic power plant designed for spacecraft by Electro-Optical Systems, Inc.

In addition to its work with liquid-fuel thermoelectric converters, Westinghouse has also built units for solar power, using an eight-foot reflector to focus sunlight on a bank of thirty-two thermocouples. Such a plant can develop 125 watts at about four volts, and Westinghouse estimated the cost of producing electric power at from seven to ten cents per kilowatt-hour, making such a system attractive even in areas less remote than space.

A distinct advantage of thermoelectricity over the solar battery is that hundreds of firms are working with thermoelectric materials, compared with the handful producing solar batteries. The reason is that nearly any heat source will operate a thermoelectric junction, while the same is not true for the solar battery.

Thermionic Conversion

A third method of changing solar energy directly to electricity is that of thermionic conversion. This involves the emission of electrons from a cathode to an anode, and the device is generally referred to as a "thermionic diode." Thermionic emission is the basis of operation of most vacuum tubes used in electronics.

Thermionic conversion is the most complex of the three methods; it is also the newest. The phenomenon was noticed by Edison years ago in his incandescent light bulb, but it was not until recently that thermionic converters have become available. Predicted in 1957 and developed in the following years by General Electric, the first converters were the size of a silver dollar and produced one watt of electricity.

The principle of thermionic conversion is simple enough: The application of heat to a cathode causes the emission of electrons and the generation of electric current. A major problem is the "space charge" phenomenon, a cloud of negatively charged ions that hinders the free flow of electrons from cathode to anode. There are three general methods of fighting this space charge, and thermionic converters are built as vacuum diodes, gas diodes, or magnetic triodes.

While the technology required for the thermionic converter is more demanding than either thermoelectric or photovoltaic approaches to power generation, the method offers advantages over both. Present materials and techniques show maximum efficiency of about 15 percent, but expected improvements could raise this to perhaps 40 percent. Some authorities consider it feasible to build a solar thermionic converter that will deliver a much higher ratio of power to weight than that offered by any other system. Power plants delivering as much as fifty kilowatts are felt to be practicable for space applications.

General Electric received an Air Force contract in 1960 for development of a five-hundred-watt thermionic converter using solar energy. Later the company developed the STEPS, or Solar Thermionic Electrical Power System, for the Air Force. Possible applications are listed as manned spacecraft, lunar bases, communication satellites, and smaller vehicles. The STEPS plant included an array of 105 thermionic diodes, with heat delivered by a sixteen-foot folding parabolic mirror and associated equipment for pointing the mirror constantly at the sun. Early in 1962 the STEPS converter produced electrical power from the sun at the company's facility in the desert near Phoenix, Arizona.

Exotic Sun-Power Systems

There are other power-conversion methods in addition to the three we have discussed. Solar-powered magnetohydrodynamic (MHD) generators have been suggested. In MHD power generation a plasma, or very hot gas, is driven through a magnetic field, generating electric current. Even conventional MHD plants have not yet proven to be successful, however, and how well solar energy can be exploited in this way is questionable.

Another method of changing sunlight into electricity is the photoemissive converter. This works in a manner similar to the thermionic converter, except that it is unconcentrated light, rather than heat, that causes electron flow. As photons of light strike a suitable photoelectric surface, they dislodge electrons from the cathode, which then flow to the anode, causing an electric current. Einstein's photoelectric law pointed to the photoemissive effect in 1905, and in 1916 Millikan proved the accuracy of Einstein's predictions with actual experiments. However, exploitation since then has been quite slow.

The U.S. Army Signal Research and Development Laboratories has built experimental photoemissive solar converters that prove the feasibility of the idea. While the efficiency of the test models was quite low, an analysis indicated that 2 percent converters should be competitive with solar batteries on a cost-and-weight basis. USASRDL estimated that a watt of photoemissive electricity would cost about thirty dollars, compared with about one hundred dollars for solar batteries, and that eighteen watts per pound of converter weight might be expected.

Dr. Thomas Gold of Cornell University has proposed large-area plastic sheets generating electricity through the photoemissive effect. Even at relatively low conversion efficiency such lightweight sheets are attractive, and cost figures might be very low. Gold suggests the possibility of developing a kilowatt of power for a kilogram of converter weight. This is roughly two-thirds of one horsepower per pound, a ratio which rivals the aircraft engine. And since a plastic sheet could be rolled easily for storage, there is further advantage to the system.

Using Solar Electric Power on Earth

Thus far there have been only a few token attempts at using electricity produced directly from solar energy to supplement conventional power sources on Earth. Radio Corporation of America engineers have designed and tested thermoelectric converters using metal reflectors to concentrate sunlight and produce small amounts of power for use in rural areas. Such power plants can pump water and provide electricity for lighting and communication, refrigeration, and other uses. Westinghouse is developing such thermoelectric converters, and General Electric considers that its thermionic converters for use in space could also generate economical power for terrestrial uses.

France, with the world's largest solar furnace, also built the largest thermoelectric generator. Built near Toulon jointly by the Society for the Study of Industrial Applications of Solar Energy and the African Investment Bureau, a 170-foot prototype was tested to pave the way for a proposed 700-square-foot generator. (This is about the size needed for a rooftop power plant, incidentally.)

Hamilton Standard Division of United Aircraft took a more expensive but simpler tack and built a hundred-watt power supply for the United States Army Signal Corps using solar batteries. Compact, light, and foldable for portability, such a power pack might also be of use for engineers, explorers, and others in remote areas. And Bell Telephone Laboratories long ago proved the feasibility of powering rural telephone lines with panels of solar batteries mounted atop service poles in the area of Americus, Georgia. Some solar experts believe that there is a vastly greater potential in using solar batteries to produce electricity for domestic and even industrial use. We look now at one such proposal made recently.

Sunshine Acres

Among the proponents of solar energy as an alternative power source is William R. Cherry of NASA's Goddard Space Flight Center, a man with long years of experience in developing solar power systems for spacecraft. Cherry presented a paper at the Solar Energy Society Conference in May of 1971 on the prospects of using direct solar energy conversion in large power plants. The abstract of his proposal is interesting:

> Projections of the U.S. electrical power demands over the next 30 years indicate that the U.S. could be in grave danger from power shortages, undesirable effluence and thermal pollution. A pollution free method of converting solar energy directly into electrical power using photovoltaics on the ground shows that sunlight falling on about 1 percent of the land area of the 48 states could provide the total electrical power requirements of the U.S. in the year 1990. By utilizing and further developing some NASA technology, a new source of electrical power will become available. Such a development is attractive from conservation, social, ecological, economic and political standpoints.
>
> While the cost of producing solar arrays by today's methods prohibits their use for large scale terrestrial plants, the paper suggests how the cost may become acceptable, especially as conventional fuels become scarcer and more expensive.
>
> Some of the desirable reasons for developing methods to convert solar energy to electrical power are: to conserve our fossil fuels for more

sophisticated uses than just burning, to reduce atmospheric pollution by 20 percent, to convert low productive land areas into high productive land areas, to make the U.S. less dependent upon foreign sources of energy, and to learn to utilize our most abundant inexhaustible natural resource.

Our biggest use of solar energy is that of farming for food. Cherry in effect suggests adding another kind of farming to our exploitation of the sun: farming for electric power. Using land otherwise useless, he proposes a crop that will return the "farmer" an income of $26 million per square mile over a period of twenty years; a most attractive land scheme.

Cherry points out that 90 percent of unmanned space vehicles are powered by solar energy, and that because of its reliability solar batteries get the lion's share of the applications, being preferred over thermoelectric, thermionic, and even the dynamic processes. For these reasons he feels that the solar battery should likewise be a winner in terrestrial power plants once certain economic problems are solved.

Suggesting a solar energy farm of one square mile, which could be built on otherwise useless land, Cherry points out that it could provide the electricity for eighteen thousand homes. To supply the entire United States by 1990 would require solar farms covering 31,500 square miles (1 percent of the area of the continental United States).

Although the large solar battery array for the Apollo Telescope Mount cost about $2 million per kilowatt, Cherry estimates that for a variety of reasons a terrestrial application could be made now at a cost of about fifteen thousand dollars per kilowatt. Further simplification, and use of aluminum reflectors to boost sunlight hitting the solar batteries, might reduce costs to about ten thousand dollars per kilowatt.

Ultimately, Cherry suggests, costs might go as low as fifty dollars per kilowatt with mechanized factory production of a "solar blanket" of large size. At such a figure, a square mile of solar farm would cost about $14 million. Total installation and operation for

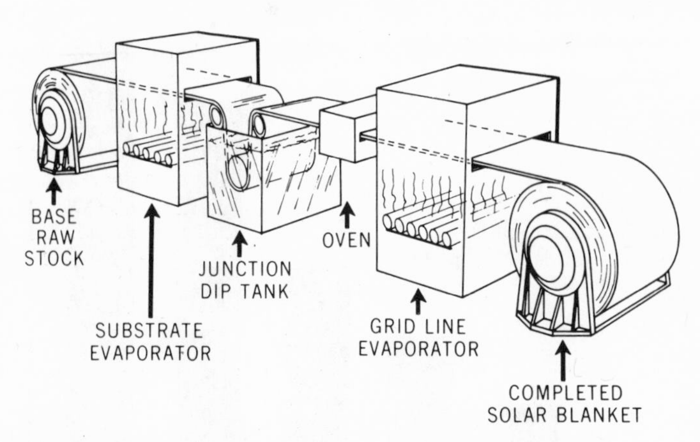

FIG. 59. Proposed method for fabricating large areas of solar cells at a very low cost.

a twenty-year period for the solar facility would amount to about $100 million. Return per acre of land through sale of electric power would be about two thousand dollars a year, a farm crop that would rate as premium.

Comparison with conventional power plants of similar size over the same period is very interesting. The cheapest electric plant now built—hydroelectric—might operate for only $170 million for twenty years. This is nearly double that of the proposed solar farm, and in some cases the hydroelectric plant could cost as much as $590 million because of location and other factors. A gas-fired electric plant would cost $463 million; oil as a fuel increases costs to $516 million. Coal is even more expensive at $534 million, and the "bargain" nuclear power plant actually costs far more at $653 million! These figures, of course, do not include the hidden costs of pollution or of radiation dangers.

Cherry's proposal obviously depends on a dramatic reduction in the cost of solar batteries. However, it is based on the modest conversion-efficiency ratio of about 7 percent for the batteries, and it is probable that with development a twofold increase in efficiency is achievable. At present only about 2 million tiny solar cells

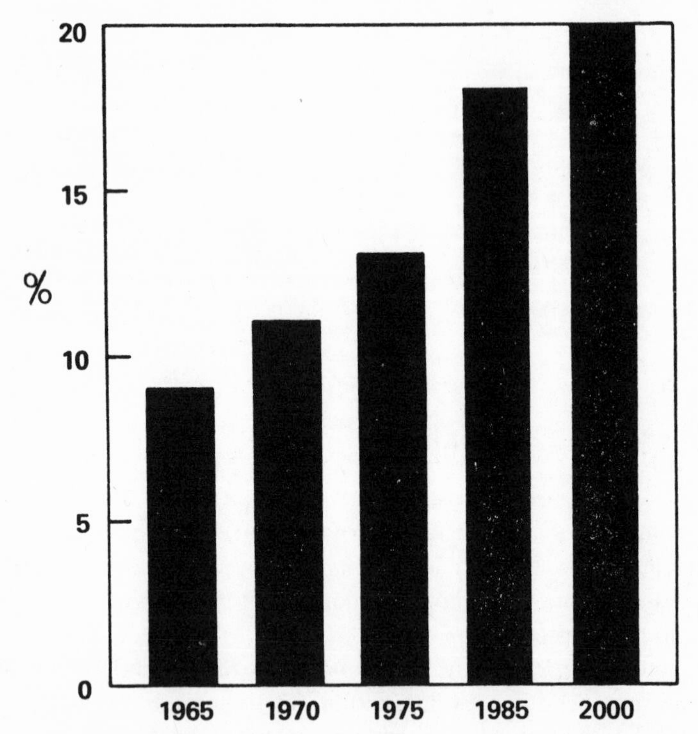

FIG. 60. Improvement of solar battery efficiency predicted with technology expansion.

are marketed a year, for a total value of not more than $8 million and on a very sporadic and inefficient basis.

For simple elegance, nothing can match the marvelous solar battery that simply sits in the sun and changes light into electric power. Conventional power plants fired with fossil fuels have a maximum efficiency of between 35 and 40 percent, achieved only with great refinements in engineering, materials selection, and so on. Maintenance costs are high, and replacement is necessary periodically as high-speed parts wear out. Waste products must be removed, some of them only at the expense of environmental pollution. Nuclear plants are even less efficient than fossil fuel plants, and while they do not produce smoke, soot, or ashes, they do create more thermal pollution than the conventional plants, plus the risks involved in radiation, both from the operating plant and from the waste material that must be handled and isolated for periods that may be many decades.

Because only a third or a little more of the heat energy in fossil fuels or nuclear fuels is converted to useful power, for each kilowatt of electricity about two kilowatts of thermal energy is dissipated into the environment. As we increase our energy consumption by 100 percent, we increase the pollution by 200 percent! Solar batteries, on the other hand, achieve about 10 percent efficiency in a simple, no-moving-parts operation. No fuel is required, no waste product is produced. There is not even any thermal pollution, for the photovoltaic process does not involve heat.

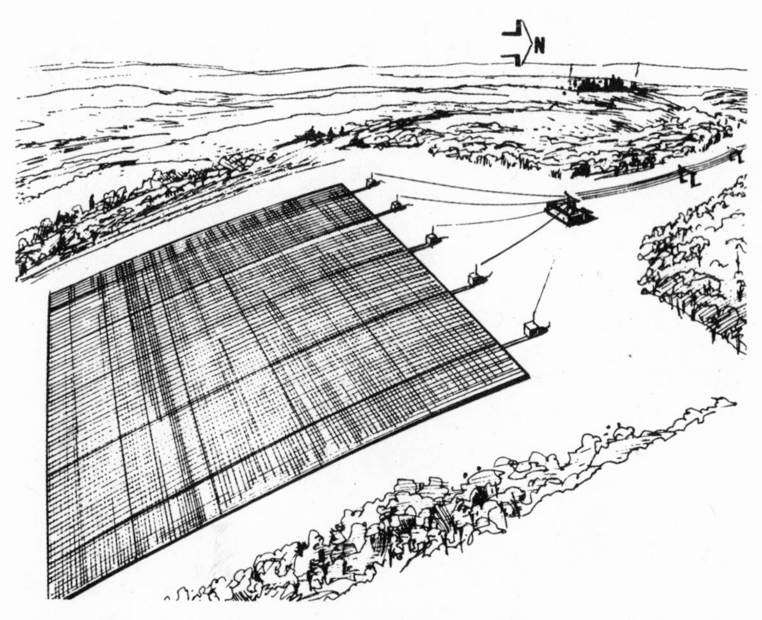

FIG. 61. A "solar farm," which would make use of cheap land to produce electricity.

It is true that solar energy is not available twenty-four hours a day or even on a regular basis, and is reduced by cloud cover and other obstructions that may block out the sun's rays. However, conventional plants are also faced with supply and demand inequities, and have to resort to power storing methods including "pumped storage," in which water is pumped uphill and allowed to fall and produce electricity when needed. This process is up to 90 percent efficient and could be used to store solar electricity as well. Cherry's proposed solar battery power plant, however, includes storage battery capacity to provide around-the-clock power.

For starters, the idea of the "solar farm" is intriguing. What other solar scientists have done with the basic concept taxes the imagination and even the belief. In the next three chapters we look at some of the suggested methods of producing electric power in great quantities from solar energy. They are ample proof that men to match the magnitude and scope of the sun's promise are among us.

In principle, solar energy conversion equipment is placed in a synchronous Earth orbit to generate electricity which is transformed into microwaves. The microwaves are then beamed to a receiving station on Earth in a spectral region in which minimum ionospheric and atmospheric absorption and scattering are encountered. The microwaves are collected at a receiving station and converted into electricity. A system consisting of a network of such satellite solar power stations could generate enough power to meet foreseeable future requirements.

Dr. Peter E. Glaser, Vice-President, Arthur D. Little, Inc.

9 THE ORBITING SOLAR POWER PLANT

In his paper on large solar-electric "farms" William Cherry also mentioned the possibility of collecting power not on the ground, but miles above the ground in outer space. Such a system has immediately obvious advantages including no losses from clouds or other obstructions, twenty-four-hour collecting ability if more than one satellite is used, plus stronger solar radiation on which to draw. In summarizing these benefits, Cherry pointed out that while an Earth-based power station to supply the whole United States would require an area of about 31,500 square miles, an orbiting station would need only 3,700 square miles, or about one-eighth as much. The idea of such an orbiting solar power plant is intriguing, and has appealed to the imagination of solar scientists for many years.

In 1965 Leon Gaucher of Texaco's Scientific Planning, Research and Technical Department, proposed in *Solar Energy* the use of an orbiting satellite to provide electric power for the United States. He said ". . . by now we are beginning to realize that this energy can be collected and concentrated with satellites and then

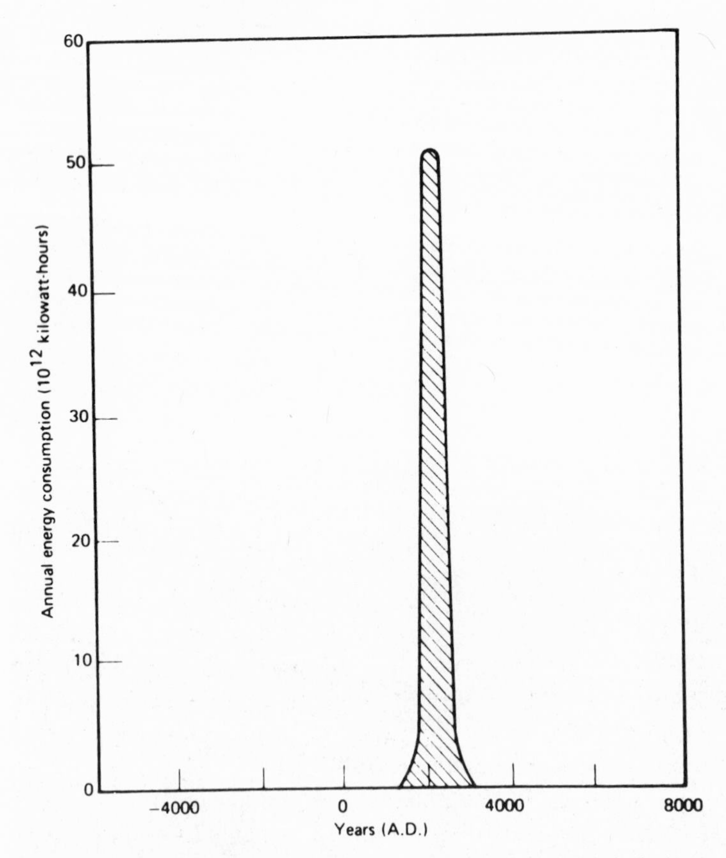

FIG. 62. Graph showing the relatively short duration of fossil fuels because of the rapidly increasing demand for power.

transmitted to the Earth in concentrated beams of selected wavelength to minimize diffusion and masking by the atmosphere." Gaucher estimated that a satellite of an area of about 462 square miles would be required to provide for our power needs in 1965. Note that Cherry's estimate for 1990, just twenty-five years later, calls for 3,700 square miles, indicative of the rapid increase in demand for power.

SSPS, the Satellite Solar Power Station

In 1968 Peter Glaser, a scientist with Arthur D. Little, Inc. (the research firm that had earlier studied solar energy exploitation in the production of algae), and president of the Solar Energy Society, made a detailed proposal for an orbiting satellite solar power station. Included in his paper, published in the May issue of *Solar Energy*, was an ominous-looking curve tracing the very narrow "pulse width" of fossil fuel availability. In 1965 the world had used about 3.3 trillion kilowatt-hours of energy. Predictions for the year 2000 showed the United States alone using 10 trillion kilowatt-

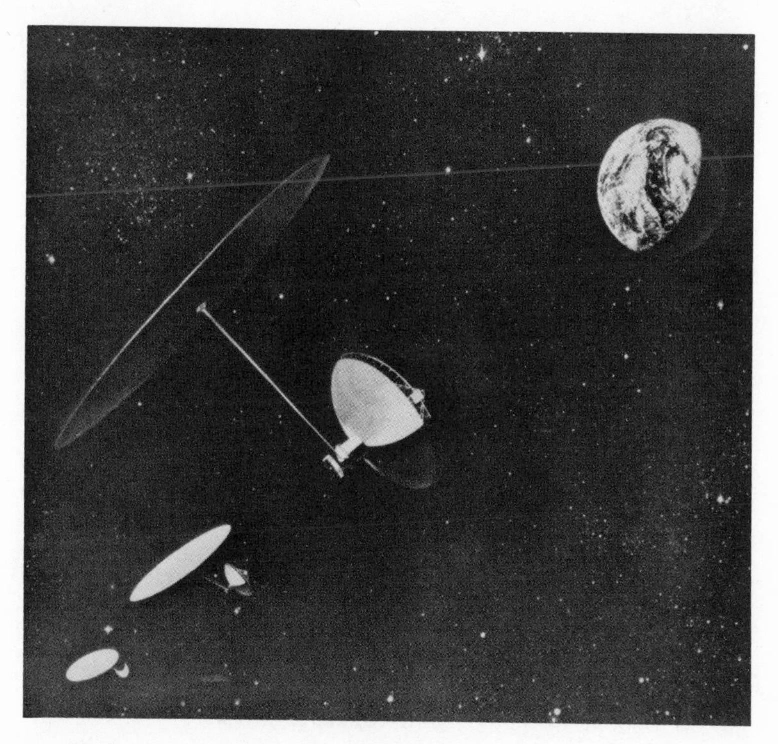

FIG. 63. An array of satellite solar power stations.

hours! The whole world would then be using 30 trillion kwh annually. We had better begin to look at some alternatives, Glaser suggested, and his preference was the orbiting satellite.

He pointed to advantages in addition to those Cherry mentioned: There would be no wind effects on collector structures, and with proper geometry and orientation of the collector it would receive the direct rays of the sun rather than a grazing angle for most of the day. To prevent losses during shadowing of the sun by the Earth, Glaser proposed two synchronous satellites spaced 21 degrees apart, an angle that would separate the satellites by about seventy-nine hundred miles. Both would be constantly above the horizon, however, and on a direct line of sight to the same point on Earth.

The Earth makes a shadow with a 17-degree angle, through which the orbiting satellite would pass for about one hour each day for twenty-five days preceding and following the equinoxes. Glaser suggested a network of satellites that could continuously serve a number of ground stations only one less than the total number of satellites. He even foresaw the use of a network of satellites serving different ground stations on a time-sharing basis. In this system a satellite might be alternately switched from one to the other of several ground stations to provide power as required.

As did Cherry, Glaser selected solar batteries as the conversion method for the satellite power station. He proposed two types of batteries. First were silicon batteries of conventional nature, yielding an efficiency of 15 percent. This approaches the maximum capability for such devices, and is twice the figure used by Cherry. However, it is in the realm of probability some years ahead. Glaser's alternative was the organic solar battery, with a potential conversion efficiency of 80 percent. These would be large-area, thin-film organic cells consisting of dyes, coatings, and plastic sheets for protection against ultraviolet light. Additional advantages beyond high conversion efficiency are those of reduced weight and cost, and the ability to be fabricated in very large areas.

Rather than one gigantic power plant of an area of several

hundred square miles, Glaser suggested smaller assemblies, each beaming its power to a region or area. As an example he pointed out that to supply the 1966 demands of the northeastern United States would require an organic solar battery satellite of only 3.3 miles in diameter. This would produce 250 million kilowatts of power, and would weigh only about 330,000 pounds, plus whatever supporting structure was needed. Since present space programs call for launching payloads as high as 100,000 pounds with a single booster, Glaser foresaw no problem in putting together such a satellite. The 3.3-mile-diameter collector would appear as large in the sky as a one-foot sphere at a distance of one mile, and pose no shadowing problems, as might be the case with a satellite twenty or thirty miles across.

Collecting the Power

Obviously the first phase of the SSPS operation is the collection of incoming solar energy. Except for the colossal size of the project, this is a simple process and is being done all the time in space. Being a static device with long lifetime and no maintenance requirements, the solar battery seems ideal for such an application.

Rather than a single unit many miles across, each collector would be made up of segments, these to be interconnected much as electrical engineers assemble panels of batteries to provide required voltages. Control of the collectors for focusing and maintenance of orbital altitude and spacing would be accomplished with gas-fueled reaction engines placed around the rims of the satellites.

A single solar battery produces only about half a volt but linking many together in proper circuits would produce hundreds of kilovolts to be transmitted on cables to the central transmitting antenna for beaming to Earth. "Super-conducting" cables would be used to minimize losses, taking advantage of the cold of space to maintain the very low temperatures required. Scope of the SSPS is indicated in the two-mile length of these cables joining the collector and transmitter.

A typical transmission-line problem would be that of carrying 100 million kilowatts (20 kilovolts at 5 million amps) through two two-inch-diameter conductors cooled to a temperature of minus 260 degrees C., each of them suitably insulated. Multiple-stage refrigerators would provide the low temperatures required, and an estimated 1,000 watts of refrigeration capacity would be sufficient. Glaser points out that both the state of the art of thermal insulation and the development of the refrigeration hardware itself are already at an advanced stage, and should be applicable to the orbiting solar power plant.

Beaming Megawatts to Earth

Collected and piped two miles to the transmitter, electric power would be ready to send on to Earth. However, it would first be converted to a form much different from either the original solar radiation or the direct electrical energy produced by the solar batteries. To minimize losses in transmission, "microwave" radiation would be used to beam power for the remainder of its journey.

Microwave radiation is more familiarly known as radar, and there is a wealth of knowledge of the handling and characteristics of this band of the electromagnetic spectrum. There is also a vast literature on experimental wireless transmission of microwave power, and several firms, including Raytheon, have conducted extensive experiments. These even include beaming power to a small helicopter flying some distance above the Earth.

Laboratory work with microwave transmission has demonstrated 425-kilowatt outputs in the ten-centimeter microwave frequency with an operating efficiency of 75 percent. Improvement to 90 percent is thought feasible. Even a 10 percent loss is appreciable when we are considering the generation of 100 million kilowatts, of course, and it is possible that the 10 million kilowatts of heat loss could be used to run auxiliary power systems aboard the satellite! Glaser suggests that space processing or manufacturing might also be accomplished on the satellite to increase its utility and minimize power production costs. With its large size the SSPS would indeed offer a roomy "factory" site.

Part of the solar power plant in the sky would be "klystron traveling wave amplifiers," which would generate the microwave radiation. Radar technology also contributes to this part of the system. Glaser suggested a microwave wavelength of ten centimeters to minimize if not completely eliminate atmospheric absorption of the transmitted energy. He proposed a transmitting antenna in the shape of a dish about two kilometers in diameter. This would be aimed at a collecting "rectenna" on Earth.

Glaser envisions ten thousand microwave generators, each producing one thousand kilowatts of power, assembled in an array about five miles on a side, feeding into a single paraboloidal antenna. Such a structure would be based on present communication large antenna technology, which generally uses petal-type or unfolding thin-shell structures. An alternative would be an array of panels rather than the single antenna for transmitting energy to Earth.

The required accuracy for facing the collector at the sun would be about 1 degree, within the present state of the art for guidance and control equipment. Accurately beaming the microwave power to Earth presents more of a problem. The transmitter should lock onto the receiving antenna and not stray more than five hundred feet in any direction both for efficient transmission of power and to safeguard the environment. This requires much greater accuracy of antenna pointing and as Glaser puts it, "stretches the present limits of attitude control techniques and pointing accuracies." Here again, experience gained in accurately pointing microwave antennas for distances of not just 22,300 miles but at the moon and even Venus, many millions of miles away, should be of benefit.

Danger: High-Energy Radiation

The receiving antenna on Earth would be made up of a large array of receivers consisting of high-efficiency solid-state rectifiers

FIG. 64. Vast banks of receiving antennas for collecting power beamed from a satellite as microwaves.

to absorb the incoming radiation from the SSPS. The microwave energy would be converted to DC power by diodes, perhaps combined in groups of four to produce a bridge-rectifier assembly. Glaser notes that gallium-arsenide diodes have already operated at tested efficiencies of 75 percent, and that 90 percent can perhaps be achieved. Again there will be the problem of a 10 percent energy loss in heat. However, Glaser believes the waste heat would be no more than that now released in a typical urban area.

Converted to alternating current, power would then be fed into a distribution network. Superconducting cables could be used on Earth as well as in the orbiting collector, to minimize losses in getting the power to domestic and industrial consumers. Conventional power plant engineers have already done much investigation and research of this method of transmission as power grids become larger and power is distributed over increasing distances and in higher voltages.

While the technical aspects of the ground-based portion of the satellite power system pose no great problems, it is on Earth that the most controversial aspect of the system is evident. Hermann Oberth and others since have suggested the possibility of huge orbiting mirrors producing "death rays" from the sky. Some critics, including solar energy proponents, are concerned that the SSPS might inadvertently become such a menace.

Glaser concedes that although the power density of the microwave beam would be less than one watt per square centimeter, and the voltage gradient where the microwaves pass through the upper atmosphere would be less than one hundred volts per centimeter and should cause no ionization of the atmosphere, the power density would be greater than normal sunlight and could conceivably damage objects or harm living tissue that might enter the beam. Major destructive effects would not result, such as the burning of homes or the melting of aircraft unlucky enough to fly through, but some care would have to be taken to prevent accidents.

It was claimed during wartime that radar waves were sterilizing technicians working in them. Although there was probably great exaggeration in such claims (as a number of surprised fathers are said to have learned), the microwave health controversy is still with us, most recently in the "radar ovens" that are becoming popular. Glaser notes that reducing the beam power density to 0.01 watt per square centimeter (one-hundredth the strength mentioned earlier by him) would provide an environment not harmful to objects or living tissue exposed only for short periods. This is the level of microwave density considered allowable in the United States. France, however, has reduced permissible levels for prolonged exposure to only 0.001 watt per square centimeter. This is only one-thousandth the density first mentioned by Glaser, and would necessitate a huge microwave receiver on Earth.

Miracle Microwaves

It is easy to be suspicious of the whole concept of the satellite solar power plant, particularly when one reads accounts of the lack of danger from the transmitted microwaves since the power reaching the ground would be "less than that of normal sunlight." If this were true the entire project would be a colossal waste of time and effort and money. However, even accepting the fact that the power density reaching Earth from the transmitter is vastly more concentrated than sunlight, wouldn't it really be simpler just to collect the same amount of power with a ground-based collector? Glaser says no for several reasons. First, the power density in space is far higher than that encountered within the atmosphere. Add to this advantage the fact that nighttime effect is minimized on the satellite and amounts to a maximum of one hour in each twenty-four.

Even granting all these plus factors, how is it possible to get the additional power through the atmosphere without losing as much as would normally be lost to haze, clouds, and so on? Conversion to microwave frequencies is the answer. Sunlight is weakened as it passes through the atmosphere by molecules of gas that have electric and magnetic "dipole moments," and also by molecules of water in cloud, fog, rain, and water vapor. Sunlight can be nearly completely bounced back into the sky by a heavy cloud layer, but this will not happen with microwave energy. Proposed microwave frequencies are in the ten-centimeter range, far longer than visible light. For microwave energy beamed straight down to Earth from the satellite, losses due to the atmosphere would be about 1 percent. Even for microwave energy traversing the atmosphere at a 60-degree angle, only 2.5 percent would be lost. Through complete cloud cover, a condition that prevails about 10 percent of the time, and turns away nearly all sunlight, losses in microwave energy would be only about 2 percent. Even through a moderate rainfall, losses increase only to 3 percent.

Engineering A Solar Satellite

Admitting that the satellite solar power station poses formidable development tasks, Glaser nevertheless feels that it is within the present capabilities of systems engineering and requires no dramatic breakthroughs either in scientific discovery or engineering technology. He points out that the project appears even less of a technological gamble than that proposed by President Kennedy when he announced the Apollo moon landings as a ten-year goal. We beat the timetable to that goal, and the moon is now old hat. Much of the knowledge, and even some of the hardware acquired in that decade-long project, involving the work of thousands of scientists and engineers and expending many billions of dollars, could be applied to the orbiting satellite power plant.

Getting the SSPS into orbit would be quite a project. However, Glaser notes that space research programs call for the space shuttle

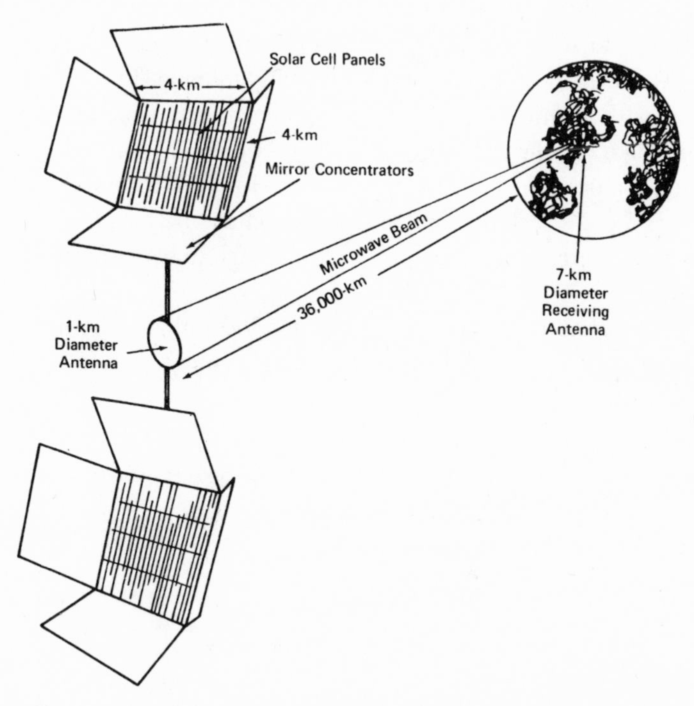

FIG. 65. Proposed pilot satellite to produce ten thousand megawatts of electric power.

going into operation in the 1975–1985 time period, and this craft is expected to reduce the cost of putting payloads into orbit to about fifty dollars a pound for low Earth orbit. After assembly of SSPS modules to produce the entire satellite power plant in this low orbit, it would be slowly moved out to synchronous altitude by a "space tug" powered by efficient but low-acceleration ion engines.

As to the feasibility of space-borne assemblies of such great size, Glaser points to careful engineering studies of even larger structures. For example, astronomers have designed radio telescopes to be assembled in space of similar structure and materials to Glaser's solar collectors. Some of the radio telescopes are even larger than the solar collectors.

One of the problems of the guidance and control mechanisms will be that of solar radiation pressure acting on the large surface of the collector and also the transmitting antenna. Even the tiny Vanguard satellite, just a few inches across, was deflected something like a mile each year it was in orbit; the orbiting solar power plant would be miles in diameter. However, techniques have already been developed and demonstrated for balancing out this disturbing force. Glaser suggests that ion engines could furnish the control.

For a satellite power plant providing 100 million kilowatts of power Glaser estimated the following weights for the structure:

Component	*Weight (Millions of pounds)*
Solar collector	2.50
Microwave antenna	1.00
Microwave generators	0.50
Generator cooling equipment	0.50
Electrical transmission line	0.20
Waste heat radiators	0.10
Crew quarters	0.05
Control thrusters and fuel	0.05
Total	4.90

The entire SSPS assembly would thus weigh about 5 million pounds, of which half would be in the solar collector. Glaser admits that this figure is optimistic, and is about ten times lighter than present solar battery weight would yield. The basis he used for his estimate is a weight of about 1 million pounds per square mile of solar batteries. The microwave antenna is also about one-tenth as heavy as present technology could produce. The total weight could therefore be 40 million pounds rather than 5 million.

The Price of Solar Power

Glaser also makes estimates for the cost of the satellite power station. The collector itself would cost about three hundred dollars per kilowatt. Microwave generator, microwave antenna, and Earth receiving antenna would each cost about fifty dollars per kilowatt, and inserting the collector into orbit would come to about the same amount. Total cost per kilowatt would then be about five hundred dollars, comparable to that of a conventional power plant.

These estimates are based on payload insertion costs of a conservative one hundred dollars per pound, rather than the hopeful figure of half that which Glaser discussed. Solar cells would be produced at a cost of about one thousand dollars per kilowatt, more than ten times as cheaply as at present. The microwave generating and receiving equipment costs are based on the present state of the art.

Total costs for a 100-million-kilowatt satellite power plant would thus be about $50 billion. While this sounds like a fantastic price to those of us who only pay a monthly utility bill, it is actually only about twice the cost of building conventional fossil fuel or nuclear plants. Once in orbit and operating it would save money since it requires no fuel, an item that in twenty years exceeds the cost of construction of conventional plants.

There are other potential savings, in that it may be feasible to beam power directly to the area where it will be used and save costly transmission systems. Glaser cites the growing practice of underground siting of lines, an item that may add 18 percent to costs of power by 1980.

Grandiose as the 100-million-kilowatt satellite solar power plant appears, Glaser moves on quickly from there. Russian scientists long ago suggested the idea of plating the moon or parts of it with solar cells to provide a gigantic power plant in the sky. Glaser suggests improving his system by moving the satellite to orbit around a planet nearer the sun, and eventually orbiting the sun itself. The logic is simple: the energy nearer the sun is far more concentrated, and at the sun's surface is estimated to be about eighty-five thousand horsepower per square yard! Glaser abandons the microwave transmission idea, since the beam would inevitably spread out, as the sun's rays do on the journey of millions of miles. Instead, he suggests that the solar energy be converted into a form storable in some material substance, and physically transported back to Earth—somewhat like deep space mining. Back on Earth the concentrated energy would be converted as required to electricity or another form of power.

Way to Go: Straight Up

For a long time solar energy enthusiasts had been criticized for spending what time they didn't use in remeasuring the solar constant in building clever but impractical gadgets, most of them sized for back-yard hobbyists. Glaser's satellite solar power station suddenly placed the plodding science at the other extreme of the spectrum.

First proposals of the SSPS were not received with any great amount of enthusiasm. Often the reactions were those of disbelief, that no one in his right mind could propose such a fantastic project. Even among the solar fraternity there was opposition and some ridicule of the satellite idea. But the SSPS idea not only hung on, it gained in strength.

Interest was high enough among microwave engineers that the

entire December 1970 issue of *The Journal of Microwave Power* was devoted to the satellite solar power station. Author of the foreword was Ernst Stuhlinger, associate director for science at NASA's George C. Marshall Space Flight Center. Stuhlinger is the man mentioned earlier for his proposal of a solar-powered spaceship. He concluded his foreword on the SSPS in these words:

> There is no doubt that the pollution-free orbiting solar power plant, as a concept, surpasses anything, in sheer magnitude, that has been realistically planned in the past. It represents a technically feasible, but size-wise unprecedented system capable of circumventing unacceptably large amounts of environmental pollution. Unless novel, still unknown ways of producing power without pollution are discovered, man may have to turn to a system of this kind, provided that he wishes to continue to enjoy the gifts of technology as he has done during past decades.

By the spring of 1971 a study team led by Arthur D. Little, Inc., and including Grumman Aerospace Corporation, Raytheon Company, and Textron, Inc., began to explore the technical and economic feasibility of the SSPS. On February 23, 1972, the team, headed up by Arthur D. Little staffers, presented testimony on the solar satellite power plant to the Committee on Science and Astronautics of the U.S. House of Representatives. An introduction to the testimony said:

> We believe that solar energy is a viable alternative as a major power source for mankind, and that it possesses the two most desirable qualities: abundance and minimal side effects. We receive about 1 kilowatt per square meter at the surface and even this amount is subject to diurnal, weather and geographical availability. Nevertheless, the solar energy falling on the United States is 500 times the expected U.S. energy consumption in the year 2000. Worldwide, the sun provides 178 trillion kilowatts (which is 5,000 times greater than the world's geothermal capacity and 60,000 times greater than the total tidal energy).
>
> We have not, however, tapped more than a fraction of this colossal reservoir of energy, despite the fact that solar energy is available every-

where, and is free of pollution. Although we have always depended on this energy source, until recently we had thought of it primarily in terms of its natural function of providing Earth with a climate hospitable enough for life to evolve.

Obviously the team thought the time had come for bolder thinking. The Satellite Solar Power Station proposal to the House of Representatives was for a solar collector of sixteen square kilometers, a microwave-transmitting antenna of one square kilometer, and a receiving antenna on Earth of a diameter of seven kilometers. This SSPS would produce ten thousand megawatts of electric power, many thousands of times the output of the largest space power system now being developed.

Textron's Spectrolab/Heliotek Division, a specialist in solar battery research, addressed the solar battery aspect of the project. Apparently the ultralight organic solar battery suggested originally by Glaser is not yet feasible, and the proposed design used conventional silicon solar batteries. Pointing out that costs had decreased since 1957 by a factor of almost 10, and that efficiency of conversion had increased to 11 percent by 1971, Textron scientists also noted that a NASA advisory group has recommended a ten-year program for achieving a goal of 18 to 20 percent efficiency.

Also foreseen was a drop in the weight of solar battery arrays from the present fourteen pounds per kilowatt to only two pounds per kilowatt. This would be achieved by reducing the thickness of the solar batteries from the eight or ten mils minimum of today to as little as 2 mils. Lightweight mirrors would be used to concentrate sunlight on the batteries at a ratio of about three or four to one, thus further decreasing weight and cost. Textron concluded by predicting a hundredfold decrease in solar battery array costs by 1985.

Raytheon in 1963 had demonstrated the wireless transmission of electric power by beaming microwaves a distance of twenty-five feet to operate a hundred-watt electric motor. From 1964 to 1968 experiments led to the flight of a small microwave-powered helicopter, under U.S. Air Force contracts, and in 1968 NASA began to support microwave-power transmission in connection with the transfer of power between manned power stations and associated smaller satellites.

Grumman, as part of its post-Apollo planning, has been giving consideration to national needs to determine realistic future goals for space technology. These studies convinced the firm of the potential of solar energy exploited with space technology and a systems management approach. Grumman's estimate of costs for the ten-thousand-megawatt SSPS ranged from $350 per kilowatt to $2,600 per kilowatt, depending on the extent to which mass-production methods are used. The study team estimated costs more conservatively, at from $1,400 to $2,600 per kilowatt.

In its conclusion, the study team report on the satellite solar power station posed the pertinent question: "Where do we go from here?" and suggested that straight up was one good possibility:

> As indicated in the preceding sections, the technical base which has already been established permits reasonable projections of the future developments required to realize the goals of a satellite solar power station. One of the major assumptions on which the development of a system of such satellites is based is that it would have to provide a significant portion of U.S. and, eventually, world energy needs. Thus, the design and development of components will require that they be capable of mass production. This will be in sharp contrast to present techniques and the resulting costs associated with the production of space flight hardware and the existing belief that anything to be done in space has to be exceedingly expensive. . . .
>
> The twin goals of design for mass production and long operating life are a strong indication that substantial innovations in technology can be expected as work on a satellite solar power station progresses. A number of analogous advances in technology have taken place as a need for innovation was recognized. Examples are the 10 percent efficient solar cells developed in 1953; the first payload orbited in 1957, and the first transmission of microwave power in 1963. The effects of these developments on solar energy applications, long before the satellite stations are developed, would have important intermediate

benefits. Thus, mass-produced low-cost solar cells could provide power to individual households in many areas to supplement other energy sources, and micro wave power could find applications in electrical transmission networks and in improved communications satellites. The large-scale use of solar energy to generate power without pollution could sustain a highly energy-dependent world culture for much longer than the few centuries associated with fossil fuels or perhaps even nuclear power. . . .

Building a giant sunbeam collector miles across in the black, airless emptiness of space would indeed seem to be a gamble on which a nation would think carefully before betting billions of dollars. But the moon was a $20 billion gamble. And fusion is a multibillion-dollar gamble as well. Will we bank on this option for power in the future, or will it be the far different proposal detailed in the next chapter?

First of all, let me explain exactly what we have been trying to do. We have not been endeavoring to extract power from the waves, from the tide, from streams. What we had in mind, my friend Boucherot and myself, was to utilize the remarkable fact that, in tropical seas, through the paradoxical collaboration of the sun and the poles, an important and almost invariable difference of temperature is maintained between the surface sea water, which is continually heated by the sun to 75 to 85 deg. fahr., and the deep-sea water, which because of the very sluggish flow from the poles to the equator, does not rise much above the freezing point, that is, 40 to 43 deg. fahr., at a depth of 3000 ft.

Regardless of the process employed, the principle of Carnot affirms that it is possible to utilize such differences of temperature to generate power. . . .

Georges Claude, *Mechanical Engineering* (December 1930)

10 SEA THERMAL ENERGY

There is another method of converting the energy of the sun to electricity so vastly different from Glaser's proposed orbiting satellite plant as to make the two seem entirely unlike. Both are based on the idea of utilizing solar energy, but in this method water is the collector of energy, rather than solar batteries.

We have considered the uses of the sun's energy in heating water and also in making fresh water from saline water. Both these processes involve the production of heat, and heat, of course, can be converted into power. Chile pioneered in the field of solar distillation. Coincidentally, it has been almost exactly a century since American engineer Charles Wilson designed and built the world's first industrial solar still at Las Salinas, a small station on the railway from Antofagasta to Bolivia.

As the name Las Salinas indicates, the area provided only saline

FIG. 66. Solar still built in Las Salinas, Chile, more than a century ago. This facility produced about six thousand gallons of fresh water daily, consumed no fuel, and did not pollute the atmosphere.

water; if mining operations were to be carried out potable water had to be provided. This was done at first by hauling in fresh water over long distances. But the solar still, with an area of 51,200 feet, produced six thousand gallons of fresh water a day, and continued in operation for about forty years until the first freshwater pipeline was completed from the Andes down to Antofagasta.

The accompanying picture was taken in 1908, after the still had been operating for thirty-four years. Chile demonstrated the effectiveness of solar energy a century ago because there was a need for such exploitation and because the Atacama Desert region is one of the most sunshine-blessed areas in the world. Of a total possible 4,383 hours of sunshine in a year, the Atacama receives an average of 4,000. The air is also clear, and there are no dust storms as on most deserts because of the soil being held by nitrate salts.

Since the building of the freshwater pipeline, there has been no need for solar stills in the region. However, at Coya, which is not far from Las Salinas, where faint traces of the old still may still be seen, including trenches and splinters of glass, a nitrate company operates ten large solar evaporation ponds to produce nitrates. Each pond has an area of 473,600 square feet, and evaporating the water with solar energy rather than conventional fuel saves about 158 tons of oil a day. To the actual cost of the fuel must be added the transporting of it to the remote desert region. Thus the solar ponds are economically very successful, despite the fact that from an efficiency standpoint they are very minimal because of the simplicity of the design—merely a shallow pond.

It is remarkable how ideas are lost. Although there is fresh water available now in the Las Salinas region, there are many areas of

FIG. 67. Experimental solar pond in Israel for converting solar energy to power. This used the "reverse-gradient" technique to trap more of the sun's heat.

scarcity in the northern provinces of Chile. Consequently there has recently been a reawakening of interest in the potential for solar energy to minimize the cost of getting water to miners in such regions. A recent scientific paper quoted a price of fifty-seven *dollars* per thousand gallons. In the United States solar engineers are trying for a price of about fifty *cents* per thousand gallons for solar desalination.

Proposals have been made for a Chilean solar facility that would produce not only fresh water but electric power as well. This plant would be built at María Elena, a remote area blessed with clear skies but a water supply containing about forty-seven hundred parts per million of salts. The pilot plant would produce about forty-five hundred gallons of fresh water a day, plus fifty kilowatts of electric power. Cost projections indicated a price of $2.57 per thousand gallons for fresh water (as opposed to $57.00 per thousand) and 34 cents per kilowatt-hour of electricity. The latter figure is quite high because of the small size of the plant but nevertheless economically attractive in the remote María Elena area.

Twin problems can often be solved at the same time, and solar energy conversion is such a method of killing two birds with the same stone. In conventional steam-powered electrical plants water is vaporized and the steam is used to drive an engine. After producing power the steam condenses back into water and is used again. In the solar power plant distillation unit, salt water is used to produce steam, but when this is condensed as fresh water it is not recycled but drawn off as a by-product. While either part of the solar conversion operation might be too costly to be considered, the dual plant may be economically attractive.

The Solar Pond

A great deal of experimental work with "solar ponds" has been done in Israel at its National Physical Laboratory, and a clever innovation was introduced in the concept of eliminating natural convection, or movement of warm water to the top of a pond. Normally, the warmest water rises to the surface, where it is quickly evaporated. In 1948, Dr. Rudolph Bloch pointed out that a solar heat collector could be greatly improved if this convection could somehow be suppressed. Bloch had come across reports of a lake in Hungary where the reverse of normal conditions was reported; that is, the temperature of the lake water increased rather than decreased with the depth. Apparently salts in the water created a "density gradient" that prevented convection. By the end of 1958 a small pond had been built on the shores of the Dead Sea near Jerusalem. The Dead Sea is loaded with minerals, and it was an old potash works that was converted to the solar energy pond experiment.

Using concentrated brine wastes, the researchers created a pond with practically no convection. Temperatures exceeding 90 degrees C. were measured at the bottom of the pool, while the surface was approximately that of the surrounding air. Now, 90 degrees C. is 194 degrees F., not far from the boiling point of water, and very attractive as a heat source for an engine. Under the direction of Dr. Harry Tabor, director of the National Physical Laboratory, a way was found to extract hot water literally in sheets from the bottom of the pool so that it could be passed through a heat exchanger.

All of this is more sophisticated than the simple evaporation pond that nevertheless saved the Chilean nitrate company 158 tons of expensive fuel a day. Tabor has estimated overall efficiency of conversion in a solar still of better than 2 percent. While this sounds low compared to the 10 percent and better efficiency of solar batteries, an acre of pond water costs nothing like an acre of silicon! A solar pond of only one square kilometer (about one-third of a square mile) would produce about 40 million kilowatt-hours a year, the equivalent of a five thousand-kilowatt power plant. Cost estimates by Tabor show electricity produced in the five-thousand kilowatt plant for about two cents per kilowatt-hour. While this is higher than costs of large fossil fuel or nuclear plants, it would be competitive with the small diesel-powered plants usually used for such installations.

One use of the density-gradient solar pond suggested by the Israelis is the production of salt (with fresh water as a by-product). The pond would yield an efficiency much higher than conventional evaporation ponds. Used as a heat source rather than for electric power, the solar pond is even more attractive, and one square kilometer would annually yield the heat equivalent of fifty thousand tons of fuel oil.

Solar pond research in Israel has been plagued with all kinds of problems, few of them having anything to do with the basic ideas back of the scheme. One pond failed to operate when mysterious bubbles formed as the water heated. When it was drained and the soil examined it was learned that bacteria in the soil were causing gas because of the increased heat. Working on limited funds, the laboratory could not afford to dig up the soil and put down a stable bottom. Instead a rubber base was installed, but this had to be provided with venting facilities so that the bacterial gas would not balloon it up off the bottom.

Nevertheless, much valuable preliminary work has been done toward economic conversion of solar energy to power using a body of water as an inexpensive medium not only for collecting but storing this energy. In the meantime, much other work, some of it spectacular in nature, has gone on in this field.

The Man Who Tapped the Ocean

Mechanical engineers have supplied all the peoples of the earth with marvelous inventions, but we must admit that many of those inventions are imperfect. We poison rivers with effluents, charge the atmosphere with noxious fumes, jar the nerves with vibrations, deafen our ears with noises, and offend our eyes with ugly factories and a countryside destroyed by our activities. I can imagine no more beneficent work for the engineer and the scientist than the removal of all those things which still mar the great and wonderful work they have done. May all of you who are coming on to the field that we are leaving carry our work a step further toward a larger and greater efficiency than we have ever known, and remove the defects of applied science without diminishing the great benefits which it has conferred on mankind.

This is the sort of statement ecologists are making all over the world today. These particular remarks were made, not by an instant ecologist, but by Loughnan St. L. Pendred, president of the Institution of Mechanical Engineers in London on October 17 in the year 1930. In the more than forty years since its utterance little has happened to make Pendred's descriptives inappropriate; the situation has, of course, become even worse. But some men have tried very hard. Just five days after the above address was made another was given at a meeting of the Metropolitan Section of the American Society of Mechanical Engineers in New York City. Its author was Georges Claude, of Paris, France.

Claude was then sixty years old, a distinguished scientist and engineer who had already achieved fame by introducing the liquefaction of air, oxygen, and hydrogen, the commercial production of nitrogen and ammonia, and industrial uses of acetylene, neon, argon, and helium. Since 1926 he had been deeply involved in trying to extract power from the sea.

Claude noted in his paper that, although he and his colleagues had themselves conceived the idea, they later found that as early as 1882 the French scientist Jacques Arsène d'Arsonval had investigated sea thermal energy. There were also men like Campbell, an American engineer, and two Italians named Boggia and Dornig. In fact, Claude remarked that had he known that sea thermal energy was such a "beaten path" he probably would not have pursued it.

Basically the idea boils down to the Carnot principle of heat engines. A difference in temperature can be used to produce mechanical energy. And such a difference in temperature exists in the sea. Claude referred to surface temperatures in the range of 75 to 85 degrees F. compared with 40 to 43 degrees F. at a depth of three thousand feet. Here is the same principle of operation as the solar pond that we have just discussed, but with a major and vitally

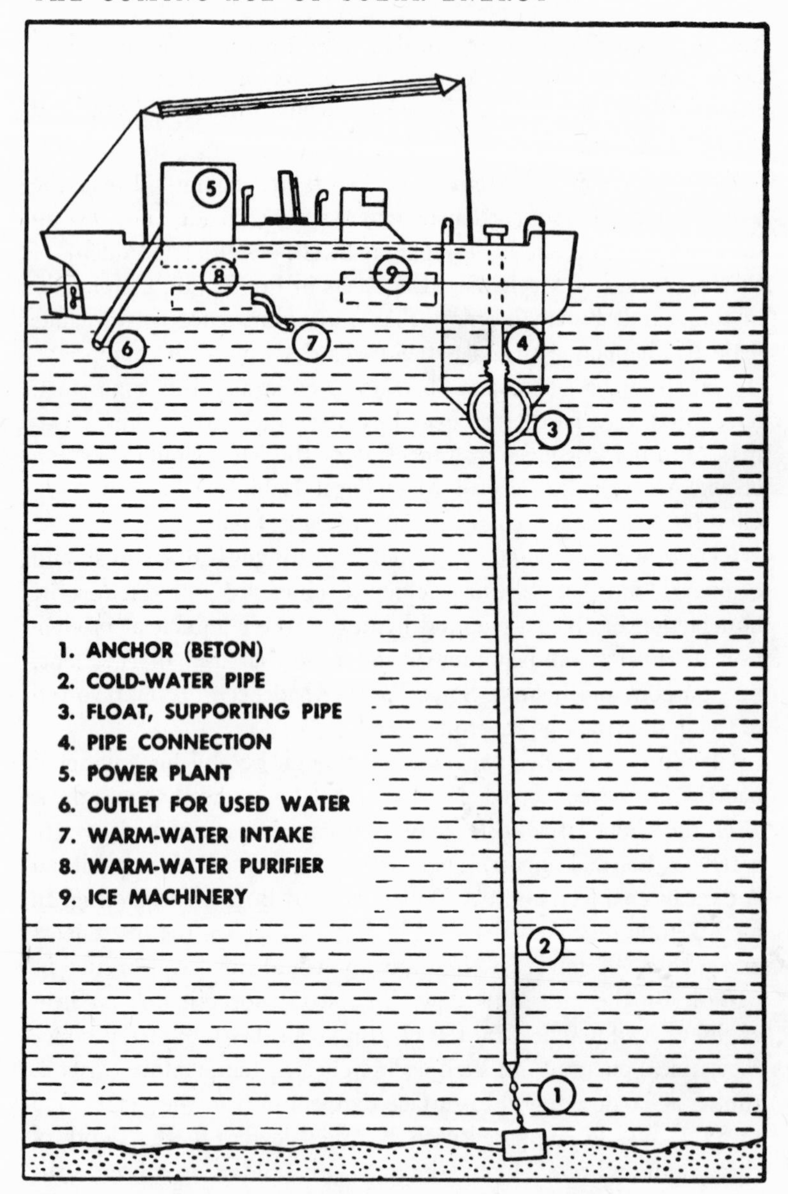

FIG. 68. Floating sea thermal energy plant, *Tunisie*.

important difference. As Israeli scientists pointed out, a solar pond must be drawn on continually, since its storage capacity is definitely limited. The sea, however, not only receives almost three-fourths of the heat from the sun, but it has been storing this heat over the ages and represents a prodigious stockpile of heat energy. There is such an abundance of energy in the sea that if man succeeded in exploiting it efficiency would be a minor factor.

Claude began his experiments in 1926 by operating in the laboratory a turbine one meter in diameter at 5,000 rpm and generating sixty kilowatts of power using a difference of only 20 degrees C. in the working fluid. This was the easy part. Next came actual tests in tropical sea waters off Cuba, which Claude picked as a good location for his sea thermal energy plant. He then went through long months of agonizing attempts to sink a steel pipe more than six feet in diameter and more than a mile long deep enough into the water to make use of the temperature difference of the sea. Plagued by hurricanes, "stupid" actions on the part of some of his crews, and other problems, Claude watched two of his gigantic water pipes sink to the bottom before he was finally successful in getting the third moored properly.

Eventually the sea thermal energy plant did work, and Claude succeeded in producing twenty-two kilowatts of power with his test engine, operating on a temperature difference of only 14 degrees C. Although some critics derisively pointed out that he was not producing as much power as he was using to pump up the water, Claude remained so convinced that he was on the right track that he proposed a pilot plant of twenty-five thousand kilowatts in the vicinity of Santiago de Cuba.

He pointed out that even this sizable venture, to cost $3 or $4 million, would not be large enough to realize his prediction of $60 per kilowatt for construction costs. That would need a plant of hundreds of thousands of kilowatts; however, as Claude said:

> But Paris was not built in a day, nor even New York. And there is already much, I think, to warrant us in stating, as I do state today, that its industries will never lack the precious energy that actuates them.

FIG. 69. Section of the intake pipe of the French experimental sea power plant at Abijan, Africa.

Claude continued his work, although the pilot plant at Santiago was never built. Instead, he suspended a cold water pipe beneath the steamer *Tunisie* to create a floating power station that again demonstrated that the STE principle would work. Claude bravely continued his costly experiments into the mid-1930s, but died without seeing any fruitful results from sea thermal energy.

Belatedly, in 1941, the French government created an organization to continue Claude's research. By 1948 a semiofficial company called Energie des Mers was formed, and some experimental work was done in North Africa toward a seven-thousand-kilowatt STE plant at Abijan. In 1948 the U.S. Department of the Interior's Office of Saline Water also poked at the fringes of sea thermal energy in a Sea Water Conversion Program at the University of California, but this effort was concerned mainly with distillation rather than power production.

STE: Sea Thermal Energy Revisited

Stimulus was given to sea thermal energy in 1961 by Asa Snyder, a vice-president of Pratt and Whitney Company, primarily a

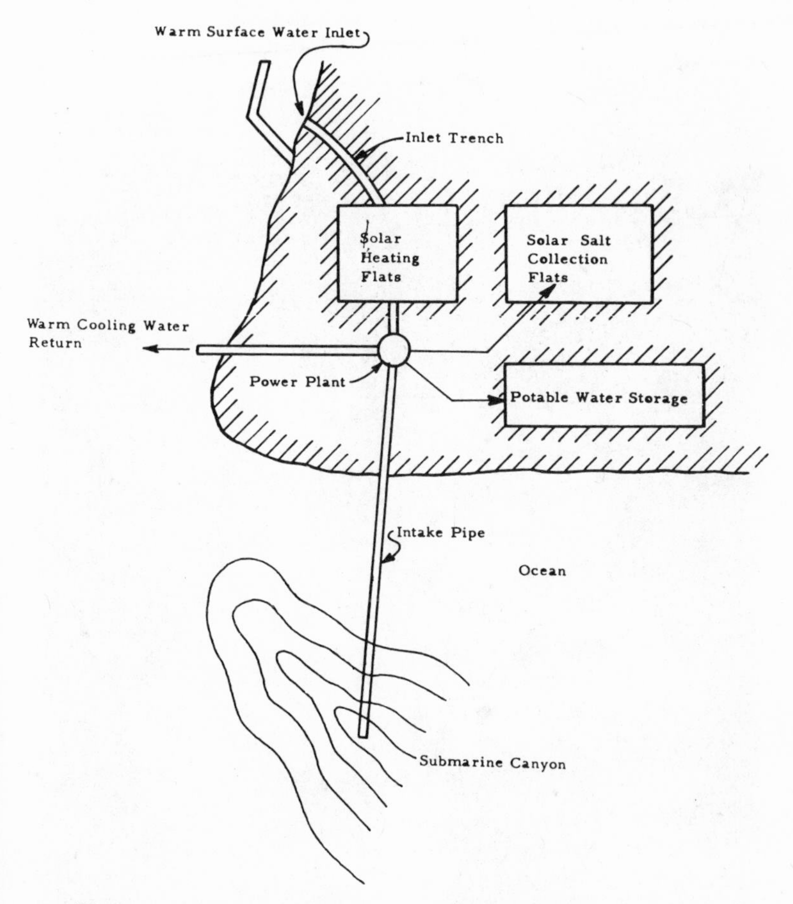

FIG. 70. Plan view of the sea thermal energy installation proposed by Asa Snyder.

builder of aircraft engines. Snyder pointed out that the potential energy in the "solar belt" between the Equator and 20 degrees north and south latitudes, where tropical waters favored the sea thermal energy idea, amounted to something like $10.5 quadrillion a year. The United States presently has a gross national product of about $1 trillion annually. Sea thermal energy "income" alone,

FIG. 71. Model of the sea thermal energy floating plant proposed by J. Hilbert Anderson.

without tapping the stored reserves, amounts to more than ten thousand times this amount. Snyder suggested that it was worth exploiting. While we could never hope to utilize all the solar energy received by the sea, Snyder suggested that nations in the "STE belt" might recover energy equivalent to more than $369 trillion annually.

Snyder proposed a modest five-thousand-kilowatt power plant combined with a 1-million-gallon-per-day freshwater plant. Erection costs would be an estimated $310 to $375 per kilowatt, competitive with conventional plants. And twenty-year projections showed even this small plant millions of dollars ahead of conventional plants because of fuel savings over the years. He felt that improvements in technology would permit far more efficient operation than Claude had attained with practically the same basic design. Surface water would be warmed by a solar heater (Claude and his associates had had little success with this method in the 1930s). The hot water would then be boiled off, using cold water brought from two thousand feet below the surface as a heat sink. Fresh water would be produced, and also salt, for an added economical incentive which Claude either did not consider or attached no importance to. Claude had operated his turbine on a temperature difference of 24 degrees F. Snyder's design called for a difference of about 50 degrees F. Surface water at 80 degrees is heated by the sun to 90 degrees. Water from two thousand feet below the surface would be at 40 degrees F.

The Floating Sea Thermal Energy Plant

Thus far we have discussed small-scale solar ponds, Georges Claude's monumental efforts, which produced about as much power as a motorcycle engine, and proposals for a five-thousand-kilowatt sea thermal energy plant. Even this last is a drop in the bucket compared with the tremendous quantities of energy waiting in the oceans of the world, and Claude pointed out more than forty years ago that power plants of hundreds of thousands of kilowatts output would be needed to economically exploit sea thermal energy. We come now to a concrete proposal for such a plant, a floating sea thermal energy plant suitable for a number of locations and producing 100 megawatts, or 100,000 kilowatts, of electric power. In addition, it would daily desalinate 60 million gallons of fresh water, produce 115 tons of oxygen plus other valuable by-products including salt and other minerals, and further enhance itself by increasing the amount of fish in the vicinity for commercial fishing! Here is a plan to exploit solar energy that grabs the imagination with a vengeance, and one that reads like a Sunday supplement science-fiction dream.

Not just dreamers, J. Hilbert Anderson and James H. Anderson, Jr., are mechanical engineers presently involved in building large-scale geothermal power plants in California. They claim that the necessary designs and hardware are possible with the present state of the art, and that the hundred-megawatt plant—which will not pollute the atmosphere or the ocean—will produce electricity at a cost of 3 mills per kilowatt-hour.

Faced with the question of Claude's failure, the Andersons point to a number of developments that make their proposed plant a far cry from Claude's pioneer effort, and will guarantee its economic as well as technological success.

The turbine would use propane gas as its working medium. This change makes possible higher efficiencies than can be had using the water itself. Propane is cheap, readily available, practically insoluble in water, and noncorrosive. The propane turbine is smaller, simpler, and cheaper to build than a steam turbine. For example, instead of a thirty-two-foot-diameter steam turbine, the Andersons have designed a propane turbine only forty-two inches in diameter but with the same power capacity.

A floating power plant will use far less pipe to reach the necessary depth than a shore-based operation with pipe slanting out to sea on the bottom. Furthermore, the floating plant can move about to ensure a constant supply of warm water. Boilers and condensers can be submerged in the floating plant, simplifying

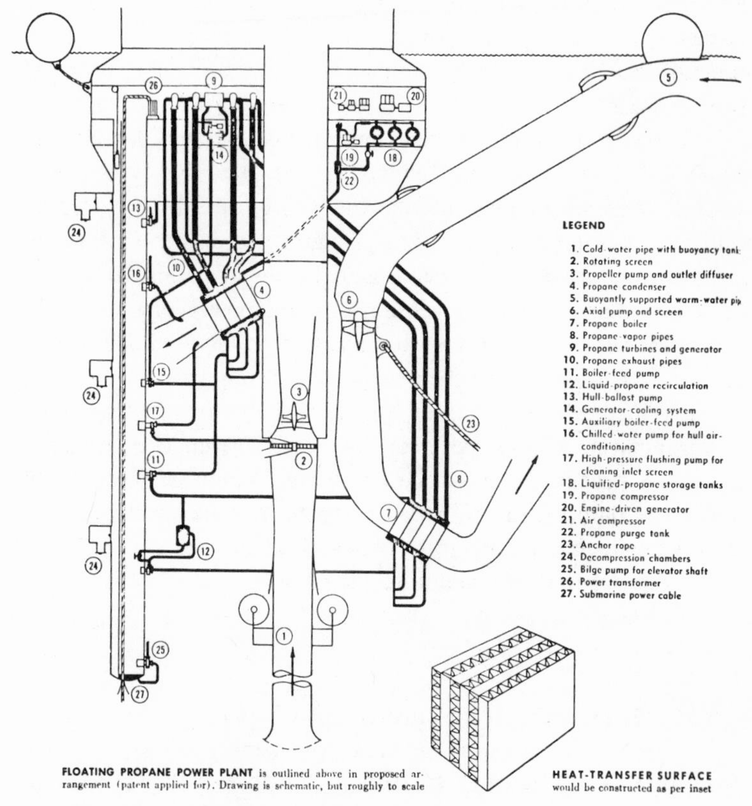

FIG. 72. Design of a floating sea thermal energy plant.

design considerations and making for better performance. With much of the structure and equipment below the surface, Anderson believes that the power plant would be stable and suffer no damage or operation problems even during heavy storms.

Anderson cites the Navy's experiences with its research vessel *Flip*, designed to be flooded to tip it into a vertical position for studies of undersea phenomena. With much of the craft's structure below the surface, it was found to be most stable and practically immune to storm damage. Work with deep-sea drilling rigs have proved that such craft can be moored accurately even in water twenty-four thousand feet deep. Also, pipes as long as six thousand feet have been suspended vertically from floating structures at sea.

The Andersons' sea thermal energy plant would be a large, bargelike structure floating freely. It would be 40 feet wide and 360 feet long, with boilers suspended 290 feet into the water, and a central cold-water pipe 40 feet in diameter and 2,000 feet long.

Operation is described as a straightforward example of the Rankine cycle used in conventional steam power plants. Warm water taken in at the surface through screens that remove foreign matter is pumped through the boilers to vaporize the propane working medium. The heated propane then drives a highspeed turbogenerator to produce electrical power. From the turbogenerator the propane goes to the condenser, where it is cooled by cold water pumped up from two thousand feet below the surface. This water is at a temperature of about 43 degrees F., compared with the surface water at about 82 degrees F.

Getting Power Ashore

A logical criticism of the floating power plant is that there is little demand for power at sea. The Andersons seriously question this argument, but for now we shall accept the fact that to the cost of producing power must be added charges for getting it to the consumer. The floating sea plant will be moored miles or even tens of miles off shore, and submarine cables will undoubtedly be used (although wireless transmission of power might be feasible, using microwaves, as Glaser has suggested for the satellite power plant).

The technology for submarine cable is available, and is neither difficult nor expensive. In fact, Sweden already transmits large blocks of electrical power in this manner. Power is presently transmitted up to hundreds of miles on land transmission lines, some of it underground, and this is a normal part of the cost of power production. Thus the floating plant sited off shore poses no particular problems of an engineering or an economic nature. While there will be problems in transmitting power, they will be nothing of the magnitude of those in orbiting satellites, where high-voltage electricity must be transmitted 2 miles in the satellite itself, and then 22,300 miles to Earth!

Fresh Water Bonus

Electrical power production alone would be a considerable feat, but the sea thermal energy plant has many other advantages. Deaerating the warm salt water used to boil the propane, boiling it in a vacuum, and then condensing it produces fresh water. The remaining brine is drained back into the sea. Of course desalination uses some of the electrical power produced, but only half the plant's output would yield 800 million gallons of fresh water a day at a cost of three to four cents per thousand gallons. Anderson compares this with estimates of a nuclear power and desalination plant proposed for the Metropolitan Water District in San Diego. This facility would produce fresh water (using one hundred megawatts of power for each 8 million gallons of water) at a cost of twenty-two cents per thousand gallons.

It is estimated that barges could haul fresh water appreciable distances for a charge of not more than five cents per thousand gallons, which, added to the production costs, would still be less than ten cents per thousand gallons and a bargain anywhere.

The Floating Industrial Plant

Peter Glaser mentioned the possibility of some manufacturing being done aboard the satellite solar power station. The Andersons make a strong case for such a development aboard an STE.

Electric power and fresh water do not nearly exhaust the potential of the sea thermal energy plant. The designers point out that it would be beneficial to deaerate the water before moving it through the boilers. This would produce 115 tons of oxygen a day, a commodity of considerable value and usefulness. The STE plant can produce power, fresh water, and oxygen, three ingredients important in the manufacture of steel. An ocean site could also be favorable for the shipping in of other raw materials, from iron ore to coke and limestone. Steel plant locations near seaports are already common. For starters, steel plants could be built near the floating sea thermal plant. The next step would be to build them *on* the plant itself.

Among the other benefits suggested are use of waste heat from the steel furnaces to generate more power or desalinate more water in the plant, and the possibility of dumping slag on the site into deep water. Pollution would be removed from land areas, and the steel plant would also have the benefit of deep docking facilities for shipping finished products.

Aluminum reduction plants would be another possibility for siting directly on the STE plant. Large amounts of electricity are needed to make aluminum, and this would be available cheaply on the spot. Furthermore, bauxite is available in quantity in the tropics, where STE plants would be built. The Andersons cite the example of Jamaica, which now ships its bauxite to Canada but which could build an STE plant locally and process its own aluminum. Some of the largest bauxite deposits are on the northeast tip of Australia, also an ideal location for an STE plant.

Fritz Haber, the German chemist who tried to recoup his country's fortunes after World War I by extracting gold from the sea, failed in this attempt, for there wasn't enough gold to make handling all that water profitable. Haber did mine the atmosphere itself for nitrogen, used during war in munitions and in peace for fertilizers; perhaps his gold scheme may work out, too. The Andersons point to production of chemicals and fertilizers on the site in STE plants. Bromine and magnesium are now extracted profitably as a specialized operation. By combining chemical and mineral extraction with the other aspects of the STE plant, many other applications may be opened up, even including the straining of gold from the sea.

Fisheries as By-products

An ace in the hole is the prospect that the sea plant would enhance commercial fishing in its area. It has long been known that cold, nutrient-rich water coming up from the lower levels creates a fertile fishing area. The Humboldt Current off Peru is an example, and when it was belatedly exploited Peru became one of the leading fishing countries of the world. As marine biologists have pointed out, the problem is getting cold water to the surface and warming it sufficiently to keep it in the top 75 meters or so of the ocean. In this way photosynthesis and food production can occur.

As we have seen, the STE plant could do this, duplicating, on a small scale, the natural upwellings that produce fishing grounds. Not only would the plant bring up cold water, but heat it so that it does remain in the top layer, where it can be effective in feeding fish. The Andersons show how increased production of fish can be worked into the power and freshwater equation, and estimate commercial fish catches approximating one-tenth the value of the power produced.

Researchers at Columbia University's Lamont-Doherty Geological Observatory at Saint Croix, the Virgin Islands, suggest an interesting variant of Anderson's STE plant. The land-based system would use the Rankine cycle in a turbine generator, and make use of cold water from twenty-five hundred feet beneath the surface. However, warm trade winds would furnish the hot part of the cycle to vaporize the working fluid, which would be a fluorocarbon used presently as a refrigerant.

The Columbia scientists also are looking at the fish farming aspect Anderson had suggested, and believe that up to fifty tons of fish per acre is a possible yield. This is something like twenty times the food crop on an acre of land, and even higher in relation to normal fish yields per acre. Fresh water would also be produced by water condensing out of the moist trade winds as they flow over the boiler tubes.

Cost of the STE Plant

The Andersons have broken down the costs of the hundred-megawatt STE plant as engineers working with large power production systems in geothermal applications. They insist that the engineering is state-of-the-art with no breakthroughs needed. Total construction and assembly costs for the sea plant would be $16,647,000, making the cost per kilowatt of installation only $166. Yearly costs would be $1,870,000, and the cost per kilowatt-hour would be a remarkable $0.00285, not quite .3 cent. The 620-megawatt nuclear power plant at Oyster Creek was estimated to cost 3.60 mills per kilowatt-hour, more than 25 percent more. This is, of course, only initial cost; fuel for the nuclear plant will be a large item each year of its operation. There are a number of other economic advantages in addition to those already mentioned:

No real estate required.
No freezing problems.
No high temperatures involved in the process.
No silting up of reservoirs.
No fuel transportation.
Power supply completely independent of weather.
No fuel shortages.
Maximum power available in summer when demand is highest.

To a world threatened with shortages of fuels and other needs, the STE plant offers production of not only oxygen but also hydrogen, both of them valuable as fuels and for other uses.

Where Sea Power?

There are many, many sites suitable for such a power supply. It has been estimated that the Gulf Stream in the Atlantic could provide sufficient power for the entire eastern half of the United States. In fact, a rough estimate of the sea thermal power potential in the Caribbean and the Gulf Stream indicates about 182 trillion kilowatt-hours annually. Projections for the United States call for about 2.8 trillion kilowatt-hours demand by 1980.

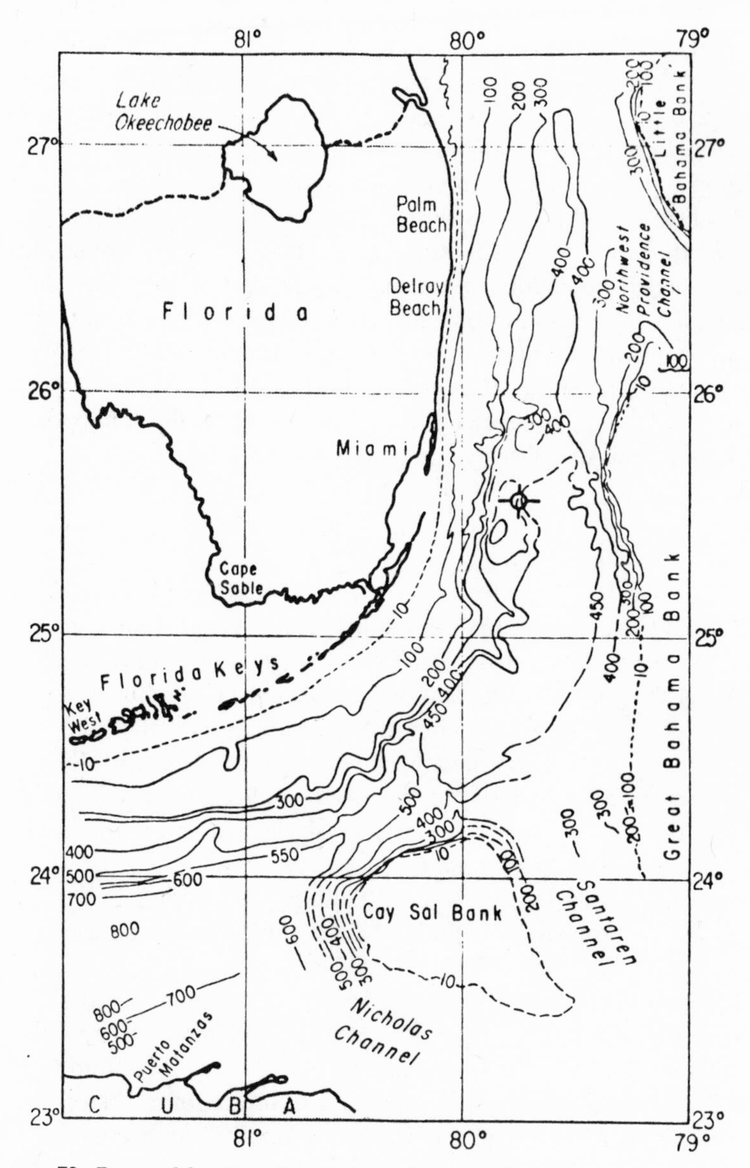

FIG. 73. Proposed location of a sea thermal energy plant for semitropical U.S. waters.

What is needed is an area of ocean where a layer of warm water overlies much colder water. This exists typically in such places as the Atlantic Gulf Stream and the Japan Current in the Pacific. Puerto Rico is ideally situated. California could count on plants located off the tip of Baja California, and these same plants could furnish power and irrigation water for much of Mexico.

There is nothing nationalistic about solar energy. At least eighteen thousand miles of coastline in the tropics and semitropics are considered useful for sea thermal energy, and at the same time good for little else. Fortunately, many undeveloped countries have ample quantities of warm water. Central America, South America, the entire Caribbean, the east and west coasts of Africa, southern Arabia, Ceylon, Indonesia, New Guinea, Borneo, the Philippines, Taiwan, and Australia are examples.

Clean Energy

One very desirable feature of a sea thermal energy plant, even beyond cheap electricity, fresh water, and a host of potential byproducts, is the fact that it will in no way pollute the environment. A conventionally fueled plant on land pumps carbon dioxide and a host of more noxious pollutants into the air, and may foul the waterways as well, both with chemicals and waste heat that is often a hazard. While a nuclear plant produces no visible pollution, the nuclear plant is an even worse thermal polluter. Nuclear radiation may also be a hazard.

Sometimes we have a choice of polluting or spending billions of dollars in eliminating pollution. Since the STE doesn't pollute, this hard choice needn't be faced. Neither does the problem of consuming our natural resources. It has been estimated that there is enough sea thermal energy to provide two hundred times the energy requirements of the entire world in the year 2000. And man will have to operate sea thermal plants for centuries before he begins to make a dent in the heat potential of all our oceans.

It is our conclusion that not only is it high time for a new look at solar energy but also that it is a key to a nonnuclear future of the world. It offers a great goal that reaches to the limits of creative imagination. We hope that this nation will recognize the immensity of the bounty to be gained from the solar harvest and give its best in this quest for solar power.

Aden and Marjorie Meinel, University of Arizona (1971)

11 A BILLION KILOWATTS OF SUNSHINE!

We have discussed two dramatic concepts for the production of larger quantities of power from solar energy: one in space and the other floating some distance off shore in tropical or semitropical waters. Now it is time to get down to earth and back to land from such concepts. Understandably, a number of solar energy proponents believe that power generated on solid ground has advantages that will more than offset those claimed for space-borne and ocean-based plants.

The only sizable mechanical solar plant ever built, the fifty-horsepower Cairo facility of Frank Shuman, was land based. Since that time no other has matched its power output, but many developments have taken place to make solar energy more attractive than it was at that time. Help comes from two major sources: the work of NASA in the development of power plants for use in space, and the efforts of independent solar scientists and engineers who have produced small solar-powered engines of various types.

The slow development of applied solar energy has been spurred by the work of John Ericsson, Dr. Charles Abbot, Frank Shuman, Willsie, and others mentioned briefly in earlier chapters. The hot-air engine invented by the Reverend Mr. Stirling long ago has been carried to a high state of development by the noted Philips firm in Holland, and also by NASA engineers working toward spacecraft power plants. Very interesting possibilities have been suggested in work on solar-powered "regenerative" hot-air engines, with efficiency close to 40 percent, some four times that of solar batteries at costs much lower for the collector.

Using parabolic reflectors, Fresnel lenses, and other concentrating devices in sizes up to forty-five feet in diameter, contemporary solar engineers have increased the output of small solar plants by a factor of 5 or so. The number of fifteen-kilowatt solar plants that

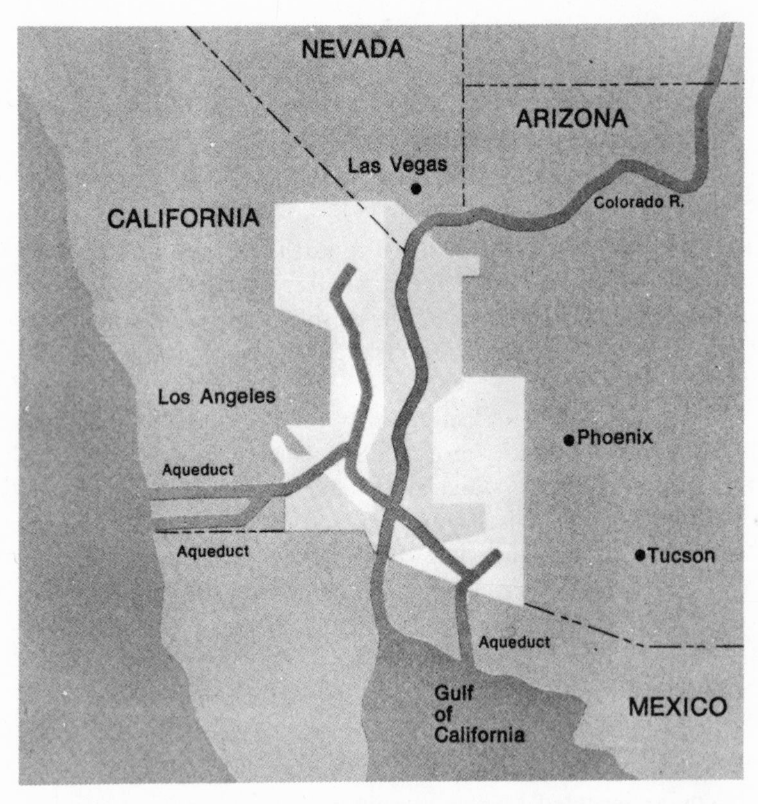

FIG. 74. Proposed area for the National Solar Power Facility. The completed complex would require about five thousand square miles of land.

have been designed and carried part way toward completion indicates that there are no great technical problems in the concept.

At Israel's National Physics Laboratory, Harry Tabor and his colleagues have built solar-powered turbogenerators producing about five horsepower. Dr. Erich Farber of the University of Florida has built small solar engines of fairly conventional design. One difference in many present-day solar power plants and those produced years ago is the use of exotic working fluids rather than steam. The Israelis, for instance, improved a small solar steam engine by using a fluid with a very high molecular weight to increase efficiency by a factor of between 2 and 3.

There has obviously been no great rush to build and use small solar power plants. Some authorities feel that interest may eventually come about in remote or undeveloped areas where a plant producing as little as 5 horsepower will have application. However, the Italian Somor organization produced solar-powered pumps of this size thirty years ago, and there was little if any demand for them. Because of this apathetic response, some solar engineers believe that solar energy will be used in appreciable amounts only when it is converted into large blocks of power, as is now done by fossil-fuel and nuclear-powered plants. It is this approach we consider now, in the proposal by two Arizona scientists, Dr. Aden Meinel and his wife, Marjorie.

NSPF: The National Solar Power Facility

On an otherwise routine day in December 1970, the *Tucson Daily Citizen* published a full-page illustrated story with the brash headline: SUN CAN POWER U.S. FOREVER. The lead paragraph stated:

> A pollution-free energy system that could supply the whole nation with all the power it needs forever could be set up in the Arizona-California desert, a University of Arizona scientist said today.

The Meinels weren't kidding, the solar fraternity soon found out. While most of their colleagues were fussing around with piddling projects, Aden and Marjorie Meinel boldly proposed a

FIG. 75. Drawing of flat-plate solar heat collectors, tipped to the proper angle for most efficiency. The power plant is shown in the distance.

vast complex in the desert bordering the Colorado River from Yuma, Arizona, to Las Vegas, Nevada. The project would be built in units each supplying one thousand megawatts, ten times the output of Anderson's sea thermal energy plant and one-tenth that of Glaser's satellite. The completed National Solar Power Facility decades down the road would produce a billion kilowatts of solar electricity, sufficient power to serve the needs of the entire United States! To get power from the Southwest all over the country, the Meinels proposed two possible methods of superconducting the electric power with little loss. One is the underground cryogenic alternating current line; the other, a superconducting direct current line.

Here was a challenge to the government and the people to put up or shut up on the twin problems of power shortages and pollution. For the solar power complex in the desert would not consume any critical fuel or materials, nor would it pollute (except to add some brine to the Gulf of California, which the Meinels considered no major problem).

In quick order the Meinels shot down conventional methods of taking care of our power needs. Nuclear power adds to the environmental problems with radiation and waste disposal. To

FIG. 76. Dr. and Mrs. Meinel, who proposed the National Solar Power Facility. Dr. Meinel holds sample of selective black coating that makes possible very high temperatures in the system.

suggestions of H-bomb electricity put forth by nuclear proponents, Meinel pointed out that this would require detonating thirty-five five-megaton hydrogen bombs in Nevada every month, a terrible price for that state to pay to provide power for the nation. In view of the booming increase in power demands, fossil fuels offered relatively short-term hope. Even the massive oil finds in Alaska would last the nation only one thousand days at the current rate of consumption, and that rate is going up every year with increased population.

The Meinels had no patience with the notion that solar energy is only for the undeveloped peoples who cannot afford "sophisticated" fuels. They also dismissed the "rooftop" concept, because while such projects were often cleverly engineered by ingenious enthusiasts, they were on far too microscopic a scale ever to succeed economically. They did not buy the orbiting satellite concept either: "Going into space to gain less than four times as much solar energy per day hardly seems worth the added expense of building the system in space." Furthermore, ". . . if the beam should be misdirected by the slightest degree it would become a tremendous death ray and just fry people."

In the Meinels' eyes, land-based solar power was the answer. The time was now, and the desert of the American Southwest was the place. The project must also be on a grand scale—and the National Solar Power Facility was nothing if not grand!

For all the science-fiction ring of the National Solar Power Facility, Aden Meinel's credentials were in perfect order. He was currently director of the sparkling new Optical Sciences Center at the University of Arizona, where such notables as astronomer Dr. Gerard Kuiper did research. A past director of the University's Steward Observatory, Meinel had been appointed to the committee that selected the Kitt Peak national observatory site. As a result, Meinel was put in charge of relocating the Steward thirty-six-inch telescope and building the new ninety-inch Kitt Peak solar telescope. For many years he had been a scientific adviser to the Secretary of the Air Force. He had also organized the Optical

Sciences Center, and had recently been elected president of the Optical Society of America.

Marjorie Meinel was deeply involved in the solar power plant design. She was a scientist in her own right, with a master's degree in astronomy. Her father was the noted Mount Wilson Observatory astronomer Edison Pettit, famous for his research on the sun, and also on the biological effects of ultraviolet light. With thermocouples he designed to measure sunspot temperatures, Pettit also measured the heat from planets and the moon. In August 1970 the International Astronomical Union honored Edison Pettit by naming a lunar feature at 28 degrees S. and 86.5 degrees W. (on the far side of the moon) Pettit Crater.

During World War II Mrs. Meinel had edited scientific journals on atomic energy and rocketry at Caltech. With her husband she had published papers on the effects of volcanic eruptions on the upper atmosphere. Somehow she also managed to raise seven children.

FIG. 77. The solar power plant, surrounded by banks of collectors.

Designing a Solar Steam Plant

William Cherry of NASA had proposed setting aside 31,500 square miles of "farmland" to harvest solar energy. The Meinels would need only 5,000 square miles because of the much higher efficiency of their conversion process. They staked out an area of uninhabited desert and in their state and neighboring California and nudging into Nevada. Here in this sunny wasteland they proposed construction of the first thousand-megawatt power plant in the desert. This would be but one-thousandth of the total, a power facility that would take an entire century to complete and to phase into the existing power production complex of fossil- and nuclear-fueled plants. According to the Meinels there would be no dislocation, no battle for survival between the solar power producers and the conventional plants. The lifetime of a conventional plant before expensive major overhauls is only about thirty-five years anyway, and the organizations presently involved in producing electricity should themselves gradually move into the production of solar power. This was so for a number of obvious reasons, and the not-so-obvious one that the turbogenerators to be used in the solar facility would be much like those in conventional plants.

Realistic about "free" solar energy, the Meinels proceeded on a hardheaded economic basis, beginning with the price of electricity in Arizona at present. Produced by natural gas, such power cost slightly more than five mills per kilowatt-hour. With that basic statistic, they studied the costs of a system designed to operate forty years (slightly longer than the lifetime of conventional plants), amortized the capital costs over fifteen years and calculated interest at 10 percent. Knowing that collector cost would be the big factor in a solar plant, the Meinels came up with an answer of about sixty dollars per square meter to build a solar plant. And they thought that price was achievable.

The Meinels felt that a solar battery conversion system as proposed by William Cherry was economically unfeasible because it presently would cost on the order of a thousand times as much as conventional power plants. Even a much more economical conversion system in which solar batteries would electrolyze water to produce gaseous hydrogen and oxygen as fuel would probably not produce even 5 percent overall efficiency.

The thermal conversion of solar energy to power generally produced low efficiencies (2 percent was the maximum cited by the Meinels) because of very low operating temperatures. Only by operating at temperatures approaching those in conventional plants could an economical conversion be made.

The Meinel solar power plant is basically a simple design. Solar heat is collected by arrays of flat-plate collectors that can be adjusted periodically for best efficiency. This heat is used to generate steam to drive a conventional steam-turbine generator, which in turn produces electrical power. Despite its basic simplicity, however, the power complex incorporates some breakthroughs that

FIG. 78. The "solartrap" method proposed by Israeli scientists at the Phoenix Solar Energy Symposium in 1955 used a selective black surface technique.

make it economically far more attractive than earlier systems. For example, the solar plant is designed for an operating temperature as high as 1,000 degrees F. (540 degrees C.).

"Selective Surface" to the Rescue

The method traditionally employed to achieve such high temperature has been the use of reflectors, mirrors, or lenses, all of which are costly to fabricate and assemble, and difficult to maintain properly focused and oriented with the sun. A Meinel innovation is the use of very efficient "selective surfaces" that absorb most of the solar heat, and radiate away very little of it. The "solar black" coating idea has been used for decades, and was demonstrated at the World Symposium in 1955 by Israeli scientists in a rooftop heat collector intended for house heating.

In two decades there had been appreciable progress with the selective surface approach. There are two basic methods of exploiting the selective surface absorption of heat. In one, Bernard Seraphin of the University of Arizona's Optical Sciences Center has developed a thin film of silicon over gold, at costs "orders of magnitude cheaper than solar batteries." A second method is the "interference film stack" originally developed by G. Hass and A. F. Turner. Such refractive thin films can be made of materials like aluminum oxide and molybdenum. While these sophisticated methods were initially prohibitively expensive, Meinel feels that a cost as low as one dollar per square meter of collector surface is now achievable.

Instead of collecting the solar heat in water, as most systems have done, Meinel proposes to use liquid sodium. Circulating between the glass collectors and a large storage tank containing a "eutectic" or soft mixture of molten salts, the sodium would transfer its heat to this storage tank, a reservoir of heat energy for the turbines.

A maximum temperature of 1,000 degrees F. is thought possible for the system, and the hotter it can be maintained the higher the efficiency of conversion of solar energy to power. However, periods of lowered solar radiation, including overnight loss of heat, resulting in reduction of temperature of the storage tank would not put the plant out of commission. The operating range would be 600 to 1,000 degrees F. The steam-turbine portion of the solar power plant would be of conventional design, and much like those used in fossil- or nuclear-fuel plants.

In designing a million-megawatt plant operating on solar energy, account must be taken of the absence of sunlight at night, and its lessened strength in the early and late hours of the day. Consequently, the plant is designed to produce a high-noon maximum not of a million megawatts but of 13 million megawatts. This will yield an average output of the designed capacity of the plant.

The Israelis had found that small steam plants could produce efficiencies of only 5 percent or less in small solar engines, and that sophisticated techniques and working fluids still could not reach 20 percent. The Meinels' power plant is expected to approach an overall efficiency of about 30 percent, including the collection and storage of heat, and its conversion to electricity.

Conventional electric power plants today approach 40 percent, and the solar facility would match this efficiency in its steam-turbine operation. In the collection of heat a remarkable 75 percent efficiency is predicted, because of the highly selective glass and semiconductor surfaces used.

It is estimated that within twenty years electricity could be coming out of the solar facilities at a cost of about five mills per kilowatt-hour, competitive with conventional power without even considering the production of fresh water or the beneficial effects of reduced environmental pollution.

Salt Water = Steam + Fresh Water

A steam-electric plant obviously requires water, and water is something the desert does not supply in copious amounts. Presently the Colorado River is the subject of great battles among the basin states for their share of its water. It will therefore be necessary to bring in sea water in huge aqueducts to provide the steam.

However, construction and pumping costs will largely be offset by the production of fresh water as a by-product.

Aqueducts to bring in salt water and to distribute fresh water as well as get rid of the brine are major construction projects that will somewhat alter the face of the desert. However, they would also open up new lands and accomplish what nothing else has succeeded in doing up to now—the process of deurbanization. In effect, a new living area will be created, and agriculture should be stimulated, too. The Imperial Valley region of southern California is already one of the richest in the world for farming. A supply of fresh water at a reasonable price should stimulate the farming economy much as the power plants will stimulate a new service community.

What About the Ecology?

Meinel admits that there are environmental prices that must be paid for building the power facility. These include production of

FIG. 79. How the desert solar power complex would look from the air, somewhat resembling a field plowed for another kind of crop.

waste heat from turbines and brine from the desalting process. Also the "fragile" desert areas would have to be altered to some extent. His answer is to disturb nature as little as possible, and to realize gains that more than offset the losses. Much desert is in effect wasteland, not even fit to support wild living things as it is. Meinel recommends a flying trip over some of the desert land to see what an arid region it presently is. While some areas are ruggedly beautiful and already are and should be preserved, much of the rest is in need of the kind of improvement that the National Solar Power Facility would bring. The U.S. Forest Service is already paving water collection areas to aid desert animals in staying alive; Meinel points out that solar collectors would concentrate what little rainfall there is on the corridors between them.

Some years ago the Atomic Energy Commission proposed building a nuclear desalination plant on the shore of the Gulf of Mexico, and there was much criticism of the plan to dump brine into the water. Meinel agrees that care must be taken, since the concentration of salt would be as high as 33 percent instead of the normal 3 percent. However, studies indicate that if brine is pumped far enough out into the Gulf, natural diffusion and the movement of the water currents will dissipate the salt sufficiently that it will create no biological or other hazard. With a project of this scope he believes it could be handled properly.

As to the net effect of removing water from the Gulf for operating the steam turbines, he points out that desert air flowing over the Gulf now evaporates about two meters of water of the Gulf a year. Using 50 billion gallons per day in the solar power complex would represent an additional two *millimeters* (a millimeter is one-thousandth of a meter) of water a year and would hardly be noticed.

While thermal pollution is an inevitable by-product of a steam power plant, the solar facility would use this heat in desalinating water. However, there have been the predictable complaints from some concerned about using up solar heat in the desert. To fears that this might drop the temperature of the desert (it cannot do so appreciably), Meinel has responded that he wishes it could, as that would make the desert more attractive to work and live in.

New Bloom for the Desert

Building a national solar power facility is a great enough undertaking to create sizable manufacturing communities. Yuma and Barstow, California, and Las Vegas would probably be the sites for such facilities. One thousand-megawatt station would require about 118 employees to operate it once it was completed. Eventually this would mean 118,000 employees. For their families and the services they would need, a large new community would be built in the desert on land opened up and made livable by the solar power complex. Probably one million jobs would be generated by the facility over the years. Only undeveloped federal lands would be used, with no private land taken up and Indian reservations and game preserves specifically excluded.

While Meinel is properly respectful of ecological considerations in the effect of the solar facility on the land, he is more concerned about the potential effects of the influx of *people* to the area as it becomes more attractive. Smog from the Los Angeles area already reaches as far as Yuma and Las Vegas on some afternoons, and this air pollution would of course have an adverse affect on the operation of the solar collectors. Thus control would have to be maintained and types of industry migrating to the source of cheap power would have to be monitored and policed. As Meinel describes it:

> The influx of people will be accentuated by the availability of power. A solar power system must be sized for production of sufficient power in January; hence there is a large surplus of "no-fuel-cost" power for the remainder of the year, except at the probable peak of air conditioning load in July, August, and September. This seasonally excess power could be used by chemical or industrial processors. For example, a jet airliner will never be plugged into a power outlet; hence liquid fuels remain a necessity for civilization. To synthesize clean-burning liquid

FIG. 80. The Meinels feel that the solar power facility would be compatible with the ecology and could be an asset to the desert wasteland selected for the site.

fuels one needs a large supply of cheap power. As a consequence, industry may tend to move into the Colorado River Metropolitan Area. Clearly, one needs the establishment of controls over the development of these desert areas. . . .

. . . Cars as we know them today would have to be prohibited from the Colorado River Metropolitan Area, but since we are talking about the first decades of the twenty-first century, gasoline cars must by then be a vanishing species since gasoline will be scarce.

Selling the Solar Farm

As late as 1969 the National Academy of Sciences, an organization with much prestige and power in the administration, dismissed solar energy as holding out little promise for contributing to the power needs of the country in the future. AEC Chairman Glenn Seaborg pointed out as he retired from that organization that state-of-the-art thermodynamic systems to be used in proposed solar energy plants were doomed at the outset by economics, since their conversion efficiency was less than 10 percent and could not hope to produce power cheaply enough to be afforded.

As have all solar enthusiasts, the Meinels found how hard it was to obtain money for research. Starting with a small grant of five thousand dollars from the University of Arizona Foundation and the National Science Foundation, they generated enough interest with their proposal to get an additional sixty-five thousand dollars from the RANN (Research Applied to National Needs) program of the National Science Foundation. The National Solar Power Facility has been described as the most rigorously researched and documented proposal ever put forward for solar energy.

By 1971 one or both Meinels were constantly on the road selling the solar energy plant idea to everyone who would listen. Their audiences even included nuclear scientists at the AEC's Livermore and Oak Ridge laboratories, and Dr. Edward David, head of the President's Office of Science and Technology.

The Meinel timetable calls for a small pilot plant of fifty-megawatt output as a demonstration of feasibility of the entire National Solar Power Facility. This would be built by 1976. The first of the thousand-megawatt plants would be completed in 1980; that is the same year, incidentally, that the Atomic Energy Commission has set for operation of its first nuclear breeder reactor, a plant that it is hoped will yield far more economical electric power.

The breeder reactor will represent about $4 billion in research and development, in addition to the $.5 billion the plant itself will cost. And to this should be added the additional billions that have been spent since the inception of nuclear-energy research. By contrast, Meinel's proposal calls for an expenditure of $10 billion over the next ten years, at the rate of $1 billion a year toward completion of the first of the thousand-megawatt units. While $1 billion is a sizable chunk of money, it represents only about one-tenth of 1 percent of the U.S. gross national product.

A prototype power plant could be built within four years, its designers estimate. It would cost about $700 million, a very attractive figure, for the Meinels pointed out the new Navajo plant in northern Arizona cost $616 million and uses coal, which costs money and also pollutes the environment in many ways, from gashing the landscape to producing air pollution and solid waste in the form of ashes.

With feasibility proved by the number-one plant, work would then begin on the other 999. As power began to be transmitted farther from the generating site (conventional plants in Arizona already send power hundreds of miles into California) cryogenic super-conducting cables would be used to minimize power losses in the transmission lines.

In presenting his ideas around the country, Meinel is often criticized for wanting to take "so much land" out of use. Generally this complaint comes from student groups who have little concept of the scale of the country. For answer, Meinel tells them that we already devote 500,000 square miles of land to the production of only 1 percent of our energy needs—the 1 percent that constitutes our food supply. There are about 130,000 square miles of desert land in the area proposed for the solar power facility, thus the use of 5,000 square miles (about 4 percent of the desert) seems to pose no major ecological or other threat. These 5,000 square miles of land needed for the project have also been compared with the 60,000 miles of grainland alone that was kept *out of production* in 1972 by the Department of Agriculture. The USDA also spent $1.8 billion in agricultural price supports in 1972 for grain.

Obviously there is sufficient land available for solar farming. There are also many unemployed scientists and engineers who would most likely be eager to accept an assignment like that of producing a new, environmentally clean power facility for the country.

With their proposed National Solar Power Facility, the Meinels have thrown down the gauntlet. They summarize their philosophy succinctly:

It is our conclusion that not only is it high time for a new look at solar energy but also that it is a key to a nonnuclear future of the world. It offers a great goal that reaches to the limits of creative imagination. We hope that this nation will recognize the immensity of the bounty to be gained from the solar harvest and give its best in this quest for solar power.

Just imagine for a moment that mankind had based his power industry on solar radiation, not fuel, and then the proposal to use different kinds of fuel was put forward. Probably there would have been very many objections. One could imagine that one of the most important arguments in defense of solar energy would be formulated as follows: Solar radiation is a "noble" form of energy and it was under its influence that life originated and continues to develop on Earth; therefore its use, no matter on what scale, could represent no danger or inconvenience for either the flora or fauna of the world. The use of any other kind of fuel would inevitably be connected with the poisoning of the atmosphere, water and land. Fuel should be used only where there are no other possibilities of obtaining energy, and in the sunny regions of the world the energy of the Sun should be used.

Professor Valentin Baum,
Tashkent (Russia) Heliotechnical Institute,
Address to the First World Symposium
on Solar Energy (1955)

12 SUN IN OUR FUTURE

In his June 1971 report on solar energy for the National Petroleum Council Committee on U.S. Energy Outlook, member Leon F. Gaucher expanded on the theme set forth above:

Had it not been for an abundance of fossil fuels—coal, oil, and natural gas—we might today have a "Solar Energy Economy" just as effective and efficient as our "Fossil Fuel Economy." If need had forced man to devote the phenomenal ingenuity and inventiveness which he has displayed in the past 150 years to the development of devices for the utilization of solar energy instead of fossil fuels, we might today have huge solar energy plants and complexes, similar to our oil refinery and chemical complexes, where the sun's energy would be col-

FIG. 81. Even in the nuclear age we are still polluting the environment, and will probably continue to do so for some time. This is Chicago on a "clear" day. All the clouds visible are man made.

lected, concentrated and stored to produce not only electric power but a whole host of other things.

In these complexes the technology of producing hydrogen and carbon monoxide from water and air and from plants and other carbon containing resources would have been perfected and, through the hydrogenation of the carbon monoxide—a process that is already well known—we would be producing hydrocarbons (synthetic crude oil and gaseous mixtures) from which we could derive the same petrochemicals and the same lubricants and fuels for mobile equipment that we produce today.

Because these solar complexes would have to be located in the hot sunny areas of the world—deserts and the like—we might have learned to transmit electrical energy over long distances more effectively, without wires perhaps.

We would have improved the heat pump to supplement the sun for the heating and airconditioning of houses and buildings and we would

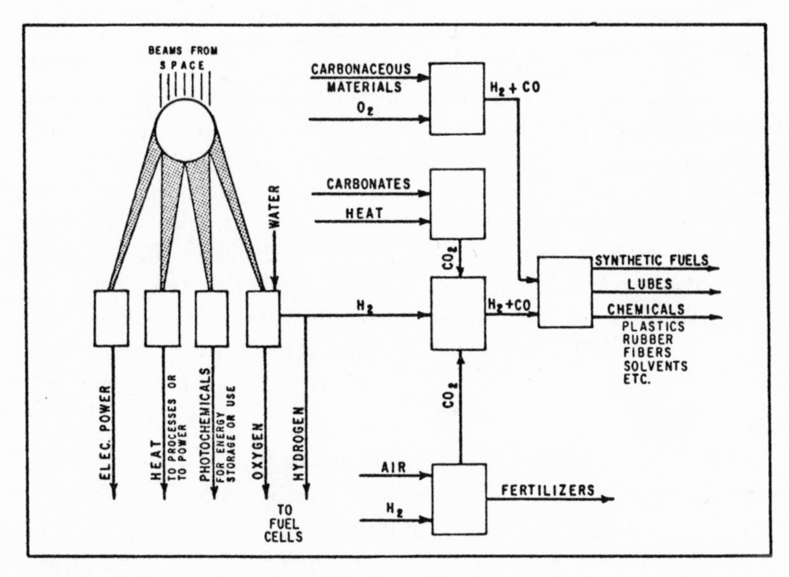

FIG. 82. Solar complex proposed by Leon Gaucher, producing not only power but heat, gases, photochemicals, and other useful products.

undoubtedly have continued the improvement of solar cells and thermoelectric and thermionic devices for the direct conversion of solar energy to electricity. Also, with improvements in the methods and cost of producing, storing, and transporting hydrogen and oxygen, we might be using fuel cells in the homes.

We most assuredly would have done a lot more on the study of photochemical reactions in which may lie the solution to the problem of storing solar energy.

Kicking the Fossil Fuel Habit

We do not have a solar energy economy, of course, and the fossil fuel economy that has brought us to the lofty plateaus of a trillion-dollar GNP has also brought us close to the brink of disaster.

Also in June 1971 President Nixon addressed the Congress on

the topic of "clean energy." The introduction was a reasoned picture of the serious situation facing us:

For most of our history, a plentiful supply of energy is something the American people have taken very much for granted. In the past twenty years alone, we have been able to double our consumption of energy without exhausting the supply. But the assumption that sufficient energy will always be readily available has been brought sharply into question within the last year. The brownouts that have affected some areas of our country, the possible shortages of fuel that were threatened last fall, the sharp increases in certain fuel prices and our growing awareness of the environmental consequences of energy production have all demonstrated that we cannot take our energy supply for granted any longer.

The bulk of Nixon's message consisted of pledges to make more nuclear energy available, and to do all possible toward cleaning up the pollution stemming from the burning of fossil fuels. At the tail end of a list of research and development goals toward clean energy, however, there was a short section titled "Solar Energy":

The sun offers an almost unlimited supply of energy if we can learn to use it economically. The National Aeronautics and Space Administration and the National Science Foundation are currently re-examining their efforts in this area and we expect to give greater attention to solar energy in the future.

Despite this commendable optimism, problems continue to plague our attempts to enjoy the twin blessings of "a high-energy civilization and a beautiful and healthy environment." Speaking to the National Petroleum Council, Secretary of the Interior Rogers C. B. Morton in February of 1972 unburdened himself of some of these knotty problems. Reciting a list of measures needed to ensure the quality of life we are seeking, he then said:

One thing the extremists in environmental groups do not understand is that these necessary jobs to enhance the quality of life can be accomplished through the use of *energy and more energy directed by an ever-escalating plateau of technology*. (Emphasis added.)

Court decisions cancelling the oil lease sale scheduled for December 21, 1971, added greatly to our difficulties, Secretary Morton protested. This setback in plans for increased oil and gas production was cause for national concern, since requirements for liquid fuels are rising much faster than our ability to supply them from domestic sources. This necessitates a large share of our fuels being purchased from foreign lands by 1980. Nevertheless, Morton put an optimistic face on the situation and stated that we would succeed in solving our power problems. Enigmatically he remarked at one point, "Let us not assume that the sun will burn out in the year 2000 and all will be over."

There is every reason to believe that the sun will indeed be burning just as brightly in the year 2000 as it has for billions of years. The question is what we will be doing about solar energy as supplement and substitute for the fossil and nuclear fuels that have got us into the bind we are in at present.

Power for the People

Solar energy is the cleanest energy we know. While the idea of successfully garnering *all* the energy in sunlight—a continuous one and one-half horsepower per square yard of area—is only a dream; even a hundredth could make us all rich with respect to power.

We may, of course, have no need of any more solar energy than what nature doles out in her own sweet time. Nuclear power may solve this worry for us by reducing our present 3.5 billions to a tiny fraction of that, most of whom will huddle in caves and forget about electric power for ages. However, being optimistic—or pessimistic depending on our views—we may look for as many as 7 billion human beings on Earth by the turn of the century.

When Columbus came to America there were only a few hundred thousand red men here. Even this relative handful, as compared with our present population density, had saturated the land. It was not capable of supporting any more lives with the technology the Indians had at their command. We will pass over the

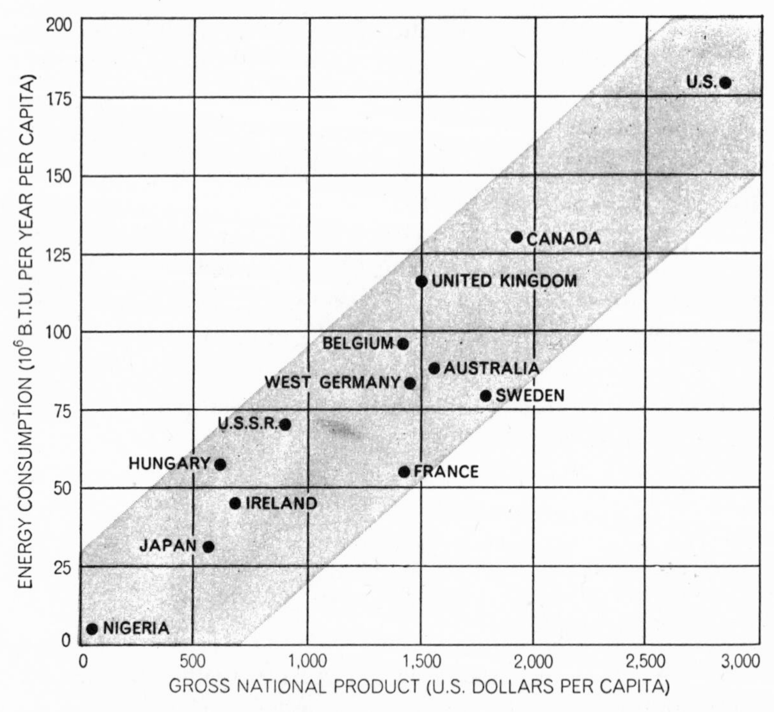

FIG. 83. The obvious correlation between gross national power and gross national product. Pollution is not plotted.

question of whether that many pastoral and contented beings are not better off than 210 million harried and hurried ones.

Generally, a higher yield of food or fuel per acre requires the expenditure of more energy. And our standard of living appears to depend on available power. With much of the world's population on short rations today it is obvious that double that number will require a much more mechanized way of life barely to stay alive. Rather than a two- or threefold increase of energy needed, the factor will in all probability be much more than that. The difficulty of assigning exact numbers is evident in the estimates now made by experts. One says that we will be using twenty times the power we use today. Another says fifty times as much.

We talk of one world in various tones, some wistfully, some scoffingly. But in nature we always see a tendency to equalize, to level things out. Temperatures and pressures try to stabilize, rather than remain at different levels. This is the case also with human beings and their demands for power, particularly as the world is shrunk drastically by increasing world trade, communications, and social and cultural mingling. We may deplore the "sucking of Eskimo pies" by Balinese dancers during intermission, but such are the facts of life. Whether or not Coca-Cola does anything uplifting has nothing to do with its wide acceptance all over the world. More to the point, the wants of the have-nots extend past sweets and soft drinks to television and refrigeration, to home lighting and mechanized travel.

Living Beyond Our Income

Of importance is the fact that while there *is* still fossil fuel left, we have already gone through the cream—if we may shift meta-

FIG. 84. Getting rid of nuclear waste uses up natural resources in the form of steel drums.

phors—and are now working on the skimmed milk, with the bottom of the bottle embarrassingly visible. Surely there is more coal and surely there is more oil. We just have to dig deeper, or haul farther, or process more. But it takes energy to mine a ton of coal, and when that energy approaches what is contained in the coal itself, we are in trouble and had better leave the stuff in the ground, where at least it will prevent a cave-in.

Just as coal and oil are limited, so are nuclear fuels. They are also difficult to process, and as the supply dwindles they become more so. Reserves of uranium and thorium used in atomic piles for the production of heat that can be converted to electricity are estimated variously as lasting from 50 to 175 years. The existence of uranium in the amount of a few grams in each ton of granite is questionable good news. Granted that this tiny amount will give a power equal to fifty tons of coal, how many tons of coal may be required to extract the grams from the ton of granite?

Many billions of dollars have been spent on nuclear energy development. Although power is being produced today, it is not possible to say that it is economically competitive as yet, considering the development costs. Fusion power, though promising theoretically, has so far been discouraging as to actual results. Even optimistic forecasts do not mention practical production of fusion power for the next twenty-five years.

Roads to Clean Energy

It has been estimated that man is using about 2 percent solar energy; that amount of his power comes from the sun in the form of wind power, waterpower, or the burning of wood and other quickly replaced hydrocarbons. As yet the use of *direct* solar energy—as with solar batteries, solar furnaces, or solar house heating—is only an infinitesimal fraction of the total consumption of energy. We cannot switch to large-scale use of direct solar energy overnight, but neither can we tolerate an energy gap that in the foreseeable future may approach 30 percent of our needs.

Starting almost from scratch, how do we begin to really put solar energy to work providing the energy for technological civilization?

In preceding chapters we have discussed three major and different proposals for making use of solar energy as a power supply. Maybe one of these will be successful, or there may be some other method that proves better. For example, in April 1972 William J. D. Escher presented a paper at a meeting of the American Chemical Society. Titled "Helios-Poseidon," Escher's paper proposed another method of achieving a sunshine economy rather than a plutonium economy, as Senator Mike Gravel expressed it when he read Escher's proposal into the *Congressional Record.* In physical appearance Escher's solar plant would resemble that of Anderson's sea thermal energy plant. It would be a free-floating complex producing electric power with a Rankine cycle turbine generator, but also using that power to dissociate oxygen and hydrogen from water by electrolysis. (Pure water would be required, of course, and this would be prepared by solar distillation.)

The electrolysis would take place at a great depth beneath the surface, for easy storage of pressurized gases. These liquid fuels could be shipped wherever needed for power conversion, or for use as valuable chemicals in a variety of processes. Hydrogen and oxygen could be used to operate fuel cells for the pollution-free generation of power, for example. Linked with the other phases of the Helios-Poseidon concept would be commercial production of seafood and the extraction of minerals. Thus, power, water, food, minerals, and fuels would be produced by this solar power complex.

Scientists at the University of Massachusetts in Amherst have also proposed a large solar power complex, including both sea thermal energy plants utilizing the thermal gradient of the sea and the production of liquid fuels by using solar energy.

Alvin F. Hildebrandt and Gregory M. Haas of the University of Houston have proposed a large solar power plant basically similar to earlier Russian concepts. The Texas scientists suggest reflecting the solar radiation from a square mile area onto a boiler atop a fifteen-hundred-foot tower. The resulting 2,000-degree K. temperature would be used to produce electricity through the MHD process. Energy would also be stored by hydrolysis of water, and an overall efficiency of 20 percent is estimated.

An imaginative solar energy application has been suggested for the Libyan Desert. Wasteland below the level of the sea would be reached by tunneling through higher ground, and filled with sea water through this tunnel. Evaporation would then constantly lower the level of this artificial lake, and thus draw water through the pipes, and this moving water would be tapped for hydroelectricity! A roundabout method to be sure, but solar energy nonetheless. And with electricity in that area selling for eight cents a kilowatt hour, and water at fourteen dollars per thousand gallons, solar energy would be not only attractive but competitive.

Dr. Farrington Daniels has suggested the creating of power or storable fuels using nothing but solar heat and the atmosphere available in the deserts of the world. In the first step water would be extracted from the air using absorbers and solar heat. The water

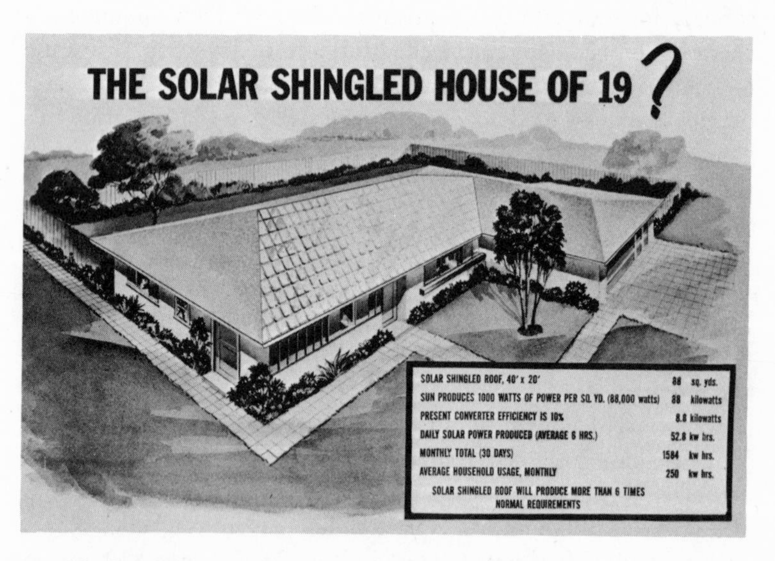

FIG. 85. Our homes may someday be self-contained power units, thanks to solar energy.

would then be broken down into hydrogen and oxygen, again by solar energy, and these gaseous fuels could be recombined in a fuel cell to make electricity or burned to produce heat or power. An alternative plan would also reclaim nitrogen from the atmosphere with a solar furnace, and mix the nitrogen with hydrogen to produce ammonia, which could be easily transported as a fuel.

Similar schemes for use on land include those in which tall vertical wind machines utilize the energy from solar-heated air rushing through them. Here again the need for expensive light-reflecting collectors is obviated. The shape of the solar energy engine of the future, then, may not be anything like the polished reflectors or glass hot boxes that have thus far taken the eye of the solar researcher. While the solar roof of semiconductor material may well take care of domestic needs, large industries may not be able to afford miles of collectors in the desert and will resort to ideas analogous to the heat pump, squeezing the energy from huge volumes of air.

Perhaps the most breathtaking solar energy concept yet is that of Russian Nobel Prize winner Nikolai Semenov, who suggested a fantastic powerhouse in the sky in which the moon is covered with semiconducting material to convert sunlight into electricity, which is somehow transmitted back to Earth!

Leon Gaucher, who suggested the solar satellite in 1965, has expanded on that idea. In his proposed solar complex, radiation received from satellites would be sorted into various quanta of energy and channeled to the conversion process best suited to their characteristics. Some radiation would go to a photoelectric converter to become electric power, some to a photochemical plant where energy-storing chemicals would be produced. Part of the radiation would be used in solar furnaces and part in solar ponds for heating and power production. Some of the energy might be used for the dissociation of water to produce hydrogen and oxygen, which could be used in fuel cells for a variety of purposes, or for chemicals, synthetic fuels, fertilizers, plastics, solvents, fibers, and the like.

FIG. 86. This central power plant would do everything conventional plants do —except burn fuel and pollute the environment.

On the other hand it may even be proved that the huge solar complex is not the only way to go, and that the more modest "solar roof" concept will have its place. Contributing to this possibility is the fact that the waste heat from electric power generation is more than sufficient to heat every home in the United States! It is also true that a good domestic oil or gas furnace is at least twice as efficient as a central power generating station.

The important point is that solar energy should be given a chance to prove its promise—or to fall on its face as the great white hope its proponents claim it to be. A gamble? Surely, but so was the man-on-the-moon project. We won that one, for what it was worth. Winning our bet on the sun would yield far more to man-on-the-Earth, both environmentally and energy-wise. There is some encouragement for solar energy proponents in the recent awarding of a contract for $446,600 by the National Science Foundation (largest solar grant to date!) for work on conversion of solar energy to power, with a goal of 20 percent efficiency. But this token effort is a tiny step on a long road.

Sun in Our Future

Man, since the time of James Watt, has found himself in the position of a "farmer" who has forgotten how to grow things, but merely goes to a cave and unearths tins of food put up by someone else long ago. If nature had not left us a legacy of coal and oil, we would today be drawing energy full-time from the sun.

The sun's radiation here on Earth is not like another kind of radiation we have become painfully aware of lately. Neither does the sun give off noxious fumes that cause horrors like the four thousand smog deaths in London in 1952. The sun brings life, and not death.

It may well be a blessing that man is at last being forced to make use of solar energy directly. As we stand on the threshold of the age of solar energy, a new age of plenty made possible by this most "noble" form of energy, how fortunate it is that "none are hid from his heat."

FURTHER READING

Abetti, Georgio. *The Sun*. New York: Macmillan, 1957.

Daniels, Farrington. *Direct Use of the Sun's Energy*. New Haven: Yale University Press, 1964.

Daniels, Farrington, and Duffie, John. *Solar Energy Research*. Madison: University of Wisconsin Press, 1955.

Gamow, George. *The Birth and Death of the Sun*. New York: New American Library, 1955.

Kiepenheuer, Karl. *The Sun*. Ann Arbor: University of Michigan Press, 1959.

Rau, Hans. *Solar Energy*. New York: Macmillan, 1964.

Ubbelohde, A. R. *Man and Energy*. Baltimore: Pelican, 1963.

ILLUSTRATION CREDITS

American Bosch Arma: Figs. 38, 41
American Telephone and Telegraph Company: Fig. 36
J. Hilbert Anderson: Figs. 71, 73
Atomic Energy Commission: Figs. 6, 81, 84
Atomics International: Fig. 5
Bell Telephone Laboratories: Fig. 25
Bridgers and Paxton: Fig. 45
William R. Cherry: Figs. 59, 61
Consolidated Vultee Aircraft Corporation: Fig. 40
Electro-Optical Systems, Inc.: Fig. 58
Energie de Mers: Fig. 69
Leon P. Gaucher: Figs. 13, 82
General Electric Company: Fig. 57
Samuel Glasstone, from *Sourcebook on the Space Sciences* (Van Nostrand, 1965): Fig. 9
Harold Hay: Fig. 47
Hoffman Electronics Corporation: Fig. 85
International Rectifier Corporation: Figs. 50, 56
Karl Kiepenheur, from *The Sun* (University of Chicago Press, 1959): Fig. 10
Laboratoire de L'Energie Solaire: Figs. 29, 44
Willy Ley, from *Engineers' Dreams* (Viking, 1964): Fig. 68
Arthur D. Little, Inc.: Figs. 60, 62, 63, 64, 65
NASA: Figs. 33, 34, 35
National Oceanic & Atmospheric Administration: Figs. 1, 3
The National Physical Laboratory of Israel: Figs. 30, 67
Nippon Electric Company: Fig. 49
Philadelphia Electric Company: Fig. 7
Power: Fig. 72
Walter Orr Roberts: Fig. 8

Scientific American: Figs. 18, 83 (September 1971)
Smithsonian Institution: Fig. 19
Solar Energy Society: Figs. 11, 15, 17, 20, 21, 22, 24, 27, 28, 31, 32, 46, 48, 55, 66, 70, 78, 86
Space Technology Laboratories, Inc.: Fig. 37
Standard Oil Company of California: Fig. 4
Stanford Research Institute: Fig. 43
From *The Sun, The Sea, and Tomorrow* (Scribners, 1954): Fig. 51
James Tallon: Fig. 12
Tucson Citizen: Fig. 75
U.N. Food and Agriculture Organization: Fig. 53
U.S. Air Force: Fig. 39
U.S. Army: Fig. 26
U.S. Department of Agriculture: Fig. 2
U.S. Navy: Fig. 23
University of Arizona: Figs. 52, 54, 74, 76, 77, 79, 80

INDEX

Page numbers in **boldface** denote illustrations.

Abbot, C. G., **42**, 43, 50, 60, 189
Abu Dhabi, 127
Adams, 40–41, 43, 51
Afghanistan, 62
Africa (see *also* countries), 32, 115, 176, **176**, 188
African Investment Bureau, 142
Agency for International Development (US) (AID), 62–63, 115
agriculture, 26, **27**, 61–62, 144
  aquiculture (hydroponics), 26, 67, 123–28, **125**, **127**
  photosynthesis and, 118–23, **119**
  "sealed in plastic," 128–29
Agriculture Department (US), 54, 113, 205
air conditioning, 60, 67, 68, 73, 103–9, **106**, **108**
aircraft, solar-powered, 50
air engines, solar, 39–40, 70, 189–90
Air Force (US), 56, 59, 67, 73, 82, 87, 93, 140–41, 163
air pollution (*see also* Ecology)
  fossil fuel, **2**, 14–16, **208**, 210
  nuclear power and, 11–12, **15**, 146–47
algae cultivation
  biopower, 129–30
  food, 26, 67, 123–25, **125**
  fuels, 128
Algeria, 39–40, **61**, 68, **68**, 103
American-Saint Gobain, 38
Ancient World, 34–36, **35**
Anderson, J. Hilbert, **178**, 180–86
Anderson, James H., 180–86
Apollo projects, 144, 165
Archimedes, 34–38, **36**, 52, 101
Arrhenius, S., 89
Arizona, 44–45, 54, 192 ff.
Arizona Republican, 44
Arizona State University, 60, 73
Arizona University, 59–60, 110, 126–27, **127**, 191 ff.
Arizona University Foundation, 204
Army (US), 58, 75–77, 141–43
Asia (*see also* countries), 32, 63, 188
Association for Applied Solar Energy (AFASE), **55**, 59–60, 68–69, 72–73, 103
Astounding Science Fiction, 93
AT&T, 80–81
Atomic Energy Commission (AEC), 11, 201, 203–4
Australia, 17, 68–69, 111, 188
automobiles, 6, 7
  solar-powered, **114**, 115
Averoni, 38
Aztecs, 34

Baker electric car, **114**, 115
Bani, University of (Italy), 70
Battelle Memorial Institute, 50, 60, 72, 109–10
batteries
  biopower (organic), 60–61, 129–30, 136
  solar, 50, 51, **56**, 57–59, 60–63, 67–69, 75–84, **76**, **78**, **79**, 115, **135**, 135–36, 142, 145–48, **146**, 196–97
Baum, V. A., 64, 206
Becquerel, A., 51
Bell Telephone Laboratories, 57–58, 80, 142
Bible, A., 72–73
biopower, 129–30
Bjorksten Research Laboratories, 60, **110**

Bloch, R., 171
boat, solar-powered, 63, 115
Boeing, 132
Boggia, 173
boilers, solar, **63**, 64–65, 215–16
Boyle, J., Jr., 45–46
Boys, C. V., 47, 69–70
Brace Experiment Station (Barbados, W.I.), 69
breeder reactor, **12**, 12–13, 204
Buffon, G., 36–38
"bug battery," 129–30
Burma, 69
Bush, V., 60

Cabot, G. L., 60
California, 43, 44–46, 54–55, 109, 113, 183, 196 ff.
California, University of, 60–61, 93, 110, 176
Cambridge Research Center (USAF), 59, 67
Campbell, 173
Canada, 69
Caribbean area, 69, 175, 186–88
Carnegie Institution, 123
Ceylon, 69
Chapin, D. M., 57
Cherry, W. R., 134, 143–45, 148–52, 196–97
Chile, 43, 54, 69, 111, 167–70, **168**
China, 69
Clarke, A. C., 80
Claude, G., 50–51, 167, 173–76, 179–80
clocks, solar-powered, 70
colleges and universities, research (*see also* names), 60–62
Colorado, 54
Columbia University, 185
Comsat, 82
Congress (US), 72–73, 163–64, 209–10, 215
Conn, W., 90–91
Cook, E., 1
cookers, solar, **42**, 43, 57, 62, 67, 69–70, **71**, 98–100, **99**
cooling, see Air conditioning; Refrigeration
cryogenic transmission, 193
Cyrano de Bergerac, 90, 91

Daniels, F., 26, 62, 121, 123, 216–17
Dannies, J. H., 70
d'Arsonval, J. A., 173
David, E., 204
Day, 51
death rays, 82, 157
Denmark, 38
density-gradient techniques, **169**, 170 ff.
deserts, utilization for solar power facilities, 191–206, 208, 216–17
developing countries (*see also* names), 62–63, 70, 97
De Caus, S., 36, 41
De Saussure, H. B., 43
Descartes, R., 36
Dornig, 173
Douglas, L. W., 59
Druids, 34

Earth Resources Technology Satellites, 97
Eastern Sun Power, 47, 48
Echo satellite, 80
ecology
  fossil fuels and, 2, **2**, **3**, 14–16, 147, **208**, 210
  National Solar Power Facility, proposed, and, 200–3
  nuclear power and, 11–12, **13**, **15**, 146–47, 193–94, **213**
  sea thermal energy and, 184, 188
  thermal ceiling, 14–16
Edison, T. A., 6, 140
Egypt, 47–48, 69–70
Egypt, Ancient, 28, 34, **35**, 36
Ehricke, K., **90**, 91
Einstein, A., 141
electricity, 6, 7
  hydroelectricity, 6, 9, 17
  magneto-hydrodynamics (MHD), 16, 141, 216
  nuclear, *see* Nuclear power
  solar (*see also* subjects), 134–65
Electro-Optical Systems, 82, **139**
electrostatic propulsion, 91–93
Eneas, A. G., 44, **45**
Energie des Mers, 51, 176
energy (*see also* specific forms; subjects)
  demand and consumption, 1–2, 5,

energy: demand and consumption (*cont'd*)
6, 7, 8–10, 14, 30–31, 210–13, **212**
resources, other than solar, 10–11, 31–33, 213–14
solar cycle in nature, 118–22, **119**
sources, historic change in, 4–7
sources, use of, 9, **30**
engines, solar (*see also* specific types), 39–51, **41**, **45**, **46**, **49**, 65, **66**, 69, 72, 189–91, 215–17
spacecraft, 84–89, **88**, **89**, **90**, **92**
England, 6, 43, 70
Eppley Foundation for Research, 60
Ericsson, J., 39, 40, 42–43, 189
Escher, W. J., 215
Europe (*see also* countries), 63–65
extractive industries, 168–70, 171, 183–84

Farber, E., 191
farms, solar, 143–48, **147**
Fink, C., 51
Finkelstein, T., 70
Fischer-Tropsch process, 128
fisheries, 185
Florence, University of, **125**
Florida, 54, 55
Florida University, 61
food supply, 8
fisheries, 185
hydroponics (aquiculture), 26, 67, 123–28, **125**, **127**
fossil fuels, 5–7, **5**, 30–31, 145, 207–10
consumption, 9–10, 30, 151–52, 194, 211
ecology and, 2, **2**, **3**, 14–16, 147, **208**, 210
exhaustion, 2, 6–7, **150**, 151, 194
photosynthesis and, 118–22, **119**, 128–29
resources, 10–11, 30–33, 213–14
solar energy comparisons, 23–25
"synthetic," 17
France, 7, 36–39, 43, 51, **64**, 65, 101–3, 108, 142, 173–76
Franklin Institute, 60, 110
French Revolution, 39
fuel cells, 16, 209, 215

Fuller, C. S., 57
furnace, solar, 34–39, **36**, **39**, 55–56, 62–63, **64**, 65, 67–69, **68**, 100–3, **102**

Galen, 34
Galileo, 36, **37**
Galvani, 129
Gaucher, L. F., 30–31, 33, 149–50, 207–9, **209**, 217
General Electric, **137**, 140–41, 142
General Motors, 60, 87
geomagnetic generators, 17
Georgia, **56**, 57
Georgia Institute of Technology, 110
geothermal energy, 17, 31–33, 71
Germany, 70
Glaser, P. E., 149, 151–64, 167, 182–83
Goddard, R. H., **49**, 49–50
Gold, T., 142
Golueke, C. G., 129–30
Government Industrial Research Institute (Nagoya), 67
Gravel, M., 215
Great Britain, 6, 43, 70
Greece, **111**, 111–12
Greece, Ancient, 34–38, **36**
Grumman Aerospace, 163, 165–66

Hass, G. M., 215–16
Haber, F., 184
Harrington, 49
Hass, G., 198
Hay, H., 107–8
H-bomb, 2–3, 74, 194
heating and ventilating, 38, 54–55, **55**, 60, 61, **61**, 67–70, 103–9, **105**, **106**, 108, 198, 208–9, **216**, 217, 218
Heidt, L., 131–32
heliograph, 52
Helios-Poseidon concept, 215
Heliotechnical Laboratory (Tashkent), 64
Heliotek, 58
Herschel, W., 20, 43
Heywood, H., 70
Hildebrandt, A. F., 213–16
Hill, R., 130, **131**
Hill reaction, 130, **131**

Hodges, C., 110, **127**
Hoffman Electronics, 58
Holland, 190
Hubbert, M. K., 10
Hughes Aircraft, 81–82
Humphrey, H., 73
Hungary, 171
Hydroculture, Inc., 127–28
hydroponics (aquiculture), 26, 67, 123–28, **125**, **127**
hydropower 6, 9, 17, 32–33, 145, 216

Iceland, 17
India, 10, 43, 62, 70, **71**, 108
Institute of Technology (Israel), 65–67
Interior Department (US), 54, 109–10, 176
International Cooperation Administration, 62
International Power Conference (Washington), 50
International Rectifier, 58, 115
Ioffe, A., 65, 138
ion (electrostatic) engines, 91–93
Israel, 55, 65–67, **66**, 70, **169**, 170–2, 191, **197**, 199
Italy, 17, 38, 70, **125**, 191

Jackson, R., 113
Japan, 55, 67, **112**, 113, 123–25
Journal of Microwave Power, 163
Journal of Solar Energy, 60
Journal Soleil, Le, 41

Kastens, M., 59
Kettering (C. F.) Foundation, 60
Khanna, M. L., 70
Kircher, 36
Kobayashi, 113
Kobayashi Institute of Physical Research, 67
Krasnovsky, 130
Kuiper, G., 194

Lange, B., 51
Langley, S. P., 43
Las Salinas (Chile), 43, 167–70, **168**
Lavoisier, A., 38–39, **39**
Lebanon, 70, 127
Libya, 216
Little (A.D.), Inc., 60, 123–24, 151 ff.
Livermore, 204
living standards, power consumption and, 6, 8–10, 211–13, **212**
Lockheed Missiles and Space Division, 93
Löf, G., 38, 54, 99, 110
Lucian, 90

magneto-hydrodynamics (MHD), 16, 141, 216
Mariner Venus probe, 82
Marly-la-Machine (France), 7
Mars Project, 86
Massachusetts Institute of Technology (MIT), 54, 61, 92, 110
Massachusetts University at Amherst, 215
McCracken, H., 112–13
McGill University, 69
Meinel, A. and M., 189, 191–206, **193**, **203**
Mellon Institute of Industrial Research, 60
Menzel, D., 29
Michigan State University, 61
microwave power, 154–66, **156**, **159**
military
  batteries, solar, uses of, **58**, 58–59
  satellites, 75 ff.
  signal mirrors, 52
  survival stills, 52–54
  weapons, 34–38, **36**, 52, 82, 157
Millikan, 141
Minneapolis-Honeywell Research Center, 60
Minnesota University, 62
mirrors, solar, 52, 82
moon, plan to use, 217
Morton, R. C. B., 210–11
Mouchot, A., 39–40, 43
mythology, 28, 34, **35**, 52

Napoleon III, 39
National Academy of Sciences, 203
National Aeronautics and Space Administration (NASA), 77, 81, 95, 97, 163–65, 189–90, 210
National Petroleum Council Committee on U.S. Energy Outlook, 207–10

National Physical Laboratory (Israel), 65, 170–72, 191
National Research, 60
National Science Foundation, 73, 204, 209, 218
National Solar Power Facility (NSPF), **190**, 191–206, **192**, **193**, **195**, **203**
natural gas, 6, 31
navigation aids, 58–59, 67, 115
Navy (US), 52–54, 57–58, 75–77
Nevada, 71–72, 192 ff.
New Mexico, 48–49
New Mexico Highlands University, 110
New Mexico Institute of Mining and Technology, 62
New South Wales (Australia) University, 69
Newton, I., 90, 94
New York University, 62, 110
New Zealand, 17
nitrate production, 168–70, 171
Nixon, R. M., 209–10
Noordung, 86
nuclear power, 74, 145, 210
  breeder reactor, **12**, 12–13, 204
  consumption, 9
  ecology and, 11–12, **13**, **15**, 146–47, 193–94, **213**
  exhaustibility, 12–13, 214
  fusion, 2–3, 13–14, 214
  output, 11, 33, 214
  sun as (*see also* Sun; specific subjects), 3–4, 17–18, 19–33

Oberth, H., 52, 80, 82, 157
oceans
  fisheries, 185
  minerals and chemicals, extraction of, 183–84, 215
  as power plants, *see* Sea thermal energy
Office of Saline Water (OSW), 54, 109–10, 176
oils, *see* fossil fuels; Petrochemicals
Orbiting Astronomical Observatory, 80
Oswald, W. J., 129–30
ovens, *see* Cookers, solar

Pakistan, 62
Paraguay, 62
Parker burning lens, 38
Patek Philippe, 70
Pearson, G. L., 57
Pendred, L. St. L., 173
Persia, 38
petrochemicals, **5**, 6–7, **7**, 31
  consumption, 9, 194
  hydrocarbons from water electrolysis and, 208
  synthetic fuels, 16
Pettit, E., 195
Philips Co., 190
phlogiston theory, 38
photochemistry, 61, 68, 118–33, 209
  aquiculture for food, 123–28, **125**, 127
  biopower, 129–30
  fossil fuels, short-term, 128–29
  photosynthesis, 118–22, **119**, 128–29
  storage, chemical, of solar energy, 130–32, **131**
photoemissive converters, 141–42
Pierce, J. R., 80
Pifre, A., 41
Pioneer satellite, 82, 84, **85**
plasma-pinch engine, 93
pollution, *see* Ecology
pond, solar, 65, **169**, 169–72, 175
power plants, solar (*see also* subjects), **218**
  engines, 39–51, **41**, **45**, **46**, **49**, 65, **66**, 69, 72, 189–91, 215–17
  satellites, 149–66, **151**, **156**, **159**, 194, 217
  terrestrial, 143–48, **145**, **147**, 191–205, **195**, **197**, **200**, **203**
  water thermal energy, 167–88, **174**, **176**, **177**, **178**, **181**, **187**, 215–17
Pratt and Whitney, 176–78
President's Materials Commission, 10, 54
Puerto Rico, 188
pumps, solar-powered, 40–41, **41**, 44–46, **45**, 62
Purdue University, 62

Quartermaster Corps (USA), 56

Rabinowitch, E., 131
RCA, 81, 93, 142
radios, solar-powered, **58**, 59, 62–63, 67
satellites, 75–77, 80–83, **81**
Ramsay, W., 6
Raytheon, 154, 162, 164
refrigeration (*see also* Air conditioning), 62, 64, 69–70, 108, 138
Republic Aviation, 92–93
Research Applied to National Needs (RANN), 204
Rome, Ancient, 34
roof, solar, 115–16, 194, 198, 218
Russia, *see* Soviet Union

sailing, solar, 93–95
Sandia, 93
Sargent, H. B., 59
Satellite Solar Power Station (SSPS), **151**, 151–66, **156**, **159**, 167
satellites, 75 ff.
communication, 80–83, **81**
Earth Resources Technology, 97
military, 75 ff.
observatories, **79**, 79–80
as power plants, 149–66, **151**, **156**, **159**, 194, 217
power plants for, 77, 85–95, **88**, **89**, **90**, **92**
weather, 77–79, **78**
Scientific American, 41–42
Seaborg, G., 11, 203–4
sea thermal energy (STE), 50–51, 61, 167–88, 174, 176, **177**, **178**, **181**, **187**, 215
Seebeck effect, 137–38
Semenov, N., 217
Seraphin, B., 198
Shuman, F., **46**, 46–48, 69–70, 189
Signal Corps (USA), 58, 143
Signal Engineering Laboratories (USASEL), 75–77
Signal Research and Development Laboratories (USARDL), 141–42
Sisler, F., 129
Skylab, **79**, 80
Snyder, A., 176–79, **177**
Society for Study of Industrial Application of Solar Energy, 142
SOCOM, 82
solar constant, 21–25, 73
Solar Energy, 149–50, 151
Solar Energy Society, 103, 111
Conference (Washington, 1971), 103, 143–44
Solar Thermionic Electrical Power System (STEPS), 141
Somor, 191
South Africa, 70
South America, 32, 62–63, 115, 188
Southeast Asia (*see also* countries), 32, 188
Soviet Union, 13, 32, **63**, 64–65, 75, 77, 125, 138, 217
space exploration (*see also* Satellites), 49, 50, 75–96
Space Technology Laboratories, 93
Spain, 70
SPUD, 86
Sputniks, 75, 77
Stanford Research Institute, 59–60
steam plant, solar (National Solar Power Facility), 190, 191–206, **192**, **193**, **195**, **200**, **203**
steam engines, solar, 39–41, 44–48, 65
stills, solar, 52–54, 61, 67, 69–70, 109–13, **111**, 167–70, **168**
earth-water, **112**, 113–14
Stirling-cycle engine, 50, 72, 87
Stirling, J., 39
Stirling, R., 39, 87, 189–90
storage, solar-energy, 62, 65, 67, 148
photochemical, 130–32, **131**
roof, 116
stove, *see* Cookers, solar
Stuhlinger, E., 91, 92–93, 163
Summers, C., 14, 16
Sun at Work, 60
sun, as energy source (*see also* specific subjects), 3–4, 19–33
energy comparisons, 23–25
energy cycle in nature, 118–22, **119**
energy spectrum, 19–20, **21**
eruptions (flares), **20**, 72
as indirect energy, 26–28, **27**

sun: as indirect energy (*cont'd*)
nuclear fusion, 3–4, 17–18, 20–21
radiation reception distribution, **24**, **25**, 26
solar constant, 21–25, 73
telescopic observation, 36, **37**
utilization problems, 28–30
Sundstrand Aviation, 87
Sunflower engine, 86–87
Sun Power Co., 47
Sunwater Co., 112–13
superconduction, 193
surfaces, selective absorptive, **197**, 198–99
Surinam, 63
Sweden, 182
Switzerland, 43, 70
Syncom, 82
synthetic fuels, 17

Tabor, H., 65, **66**, 67, 171, 190
Taiwan, 69, 188
Tarcici, A., 70
Targioni, 38
telescope, solar, 36, **37**
television, solar-powered, 62–63
satellites, 77–83, **78**, **81**
Telkes, M., **53**, 53–54, 60, **99**
Tellier, A., 41, 45
Telstar, 80, 81, **81**
Textron, 163–64
thermal ceiling, 14–16
thermionic converters, **139**, 140–41, 209
thermoelectricity, 65, 68, **137**, 137–39, 142, 209
Thomason, H., 54, 104–7, **106**
Thompson Ramo Wooldridge, 86–87
tidal power, 17, 33
Tiros satellites, **78**, 79
Trombe, F., 55–56, 65, 101–3, **102**
Tschirnhaus, 38
Tsu, T. C., 94
turboelectric generators, 85–88, 191, 215
Turner, A. F., 198
Tzetzes, J., 34

United Aircraft, 142–43
United Nations, 70–72, 127
Conference on New Sources of Energy (Rome, 1961), 70, 71–72, 110
United States (*see also* specific names, subjects)
energy consumption, 9

Van Bavel, C., 113
Vanguard projects, 50, 58, 75–77, **76**, 93–94, 160
Van Helmont, 121
Villette, 38
von Braun, W., 86

Warburg, O., 123
water
decomposition, photochemical, 130–32, **131**
desalination, 43, 52–54, 61, 62, 67, 69, 170, 183, 201
distillation, 8, 52–54, 61, 67, 69–70, 109–13, **111**
distillation, earth-water, **112**, 113–14, 167–70, **168**
electrolysis, 208, 215
food supply and, 26
heaters, 55, **63**, 64–68, 70, 109
oceans, brine additions to, 201
power (*see also* Hydropower; Sea thermal energy; Tidal power), 4, 6–7, 17, 31–32
thermal pollution, 12, **15**
Watt, J., 90, 219
Westinghouse, 93–94, 139, 142
weapons, 34–38, **36**, 52, 82, 157
Willsie, H. E., 45–46, 189
Wilson, C., 42, 167
windpower, 4, 17, **27**, 31–32, 71
Wisconsin, University of, 62
World Meteorological Organization, 32
World Power Conference, 69
World Symposium on Applied Solar Energy (Tucson and Phoenix, 1955), 59–60, 72, **197**, 198
Wolf, M., 16
Wright, F. L., 60

Yellott Solar Energy Laboratory, 60

74 75 10 9 8 7 6 5 4 3 2